Practical Electrical Network Automation and Communication Systems

Titles in the series

Practical Cleanrooms: Technologies and Facilities (David Conway)

Practical Data Acquisition for Instrumentation and Control Systems (John Park, Steve Mackay)

Practical Data Communications for Instrumentation and Control (John Park, Steve Mackay, Edwin Wright)

Practical Digital Signal Processing for Engineers and Technicians (Edmund Lai)

Practical Electrical Network Automation and Communication Systems (Cobus Strauss)

Practical Embedded Controllers (John Park)

Practical Fiber Optics (David Bailey, Edwin Wright)

Practical Industrial Data Networks: Design, Installation and Troubleshooting (Steve Mackay, Edwin Wright, John Park, Deon Reynders)

Practical Industrial Safety, Risk Assessment and Shutdown Systems (Dave Macdonald)

Practical Modern SCADA Protocols: DNP3, 60870.5 and Related Systems (Gordon Clarke, Deon Reynders)

Practical Radio Engineering and Telemetry for Industry (David Bailey)

Practical SCADA for Industry (David Bailey, Edwin Wright)

Practical TCP/IP and Ethernet Networking (Deon Reynders, Edwin Wright)

Practical Variable Speed Drives and Power Electronics (Malcolm Barnes)

Practical Electrical Network Automation and Communication Systems

Cobus Strauss CPEng BSc(ElecEng), BComm, SKM Engineers, Perth, Australia

AMSTERDAM • BOSTON • HEIDELBERG • LONDON • NEW YORK • OXFORD
PARIS • SAN DIEGO • SAN FRANCISCO • SINGAPORE • SYDNEY • TOKYO

Newnes is an imprint of Elsevier

Newnes
An imprint of Elsevier
Linacre House, Jordan Hill, Oxford OX2 8DP
200 Wheeler Road, Burlington, MA 01803

First published 2003

Copyright © 2003, IDC Technologies. All rights reserved

No part of this publication may be reproduced in any material form (including photocopying or storing in any medium by electronic means and whether or not transiently or incidentally to some other use of this publication) without the written permission of the copyright holder except in accordance with the provisions of the Copyright, Designs and Patents Act 1988 or under the terms of a licence issued by the Copyright Licensing Agency Ltd, 90 Tottenham Court Road, London, England W1T 4LP. Applications for the copyright holder's written permission to reproduce any part of this publication should be addressed to the publisher

British Library Cataloguing in Publication Data
A catalogue record for this book is available from the British Library

ISBN   07506 58010

For information on all Newnes publications,
visit our website at www.newnespress.com

Typeset and Edited by Vivek Mehra, Mumbai, India

Printed and bound in Great Britain

Contents

Preface		viii
1	**Introduction to power system automation**	**1**
1.1	Definition of the term	1
1.2	What is power system automation?	1
1.3	Power system automation architecture	4
1.4	Summary	5
2	**Historical development of power system automation**	**7**
2.1	Introduction	7
2.2	The electrical protection industry	7
2.3	The electronic industry	9
2.4	The switchgear industry	10
2.5	The automation industry	10
2.6	The computer industry	12
2.7	The communications industry	13
2.8	The measurement industry	13
3	**Overview of power networks**	**14**
3.1	Introduction	14
3.2	Power generation	16
3.3	Power transmission	16
3.4	Power distribution	18
3.5	Low voltage applications	18
3.6	Network studies	18
4	**Fundamentals of electrical protection**	**20**
4.1	The need for electrical protection	20
4.2	Overview of electrical faults	20
4.3	Protection components	21
4.4	Protection qualities	26
4.5	Main functions of protection relays	32
4.6	Specific applications	38
4.7	Comparision of electromechanical relays and digital relays	44
5	**Remote substation access and local intelligence**	**46**
5.1	Introduction	46
5.2	The remote terminal unit	46
5.3	Programmable logic controller	47
5.4	Protection relays	47
5.5	The intelligent electronic device (IED)	48

6 Data communications — 51

6.1	Basic requirements of communication	51
6.2	Communication topologies	52
6.3	Communication techniques	55
6.4	Media access control principles	55
6.5	The OSI model	60
6.6	Performance criteria	64

7 Communication protocols — 67

7.1	Overview	67
7.2	Distributed network protocol (DNP V3.0)	67
7.3	Modbus	78
7.4	Modbus plus	79
7.5	LonTalk	82
7.6	Ethernet	84
7.7	TCP/IP	87
7.8	PROFIBUS	93
7.9	IEC 60870-5-101	98
7.10	IEC 60870-5-103	102
7.11	UCA 2.0 (an overview)	106
7.12	Standardization	106

8 SCADA systems — 108

8.1	Definition and background	108
8.2	Requirements for the SCADA master station	109

9 Communications in power system automation — 119

9.1	Overview	119
9.2	Configuration	120
9.3	Communication requirements	121
9.4	Example of Requirements	127

10 Power system automation architectures — 128

10.1	Introduction	128
10.2	Types of power system automation architectures	128

11 Power system automation systems on the market — 136

11.1	Introduction	136
11.2	GE	136
11.3	ABB	141
11.4	SEL	143
11.5	Siemens	150
11.6	ALSTOM	156

12	Practical considerations	164
12.1	Justification	164
12.2	Basic strategies	164
12.3	Constraints	165
12.4	Competence management	165
12.5	Electrical protection	165
12.6	Suppliers	167
12.7	Power system automation and the Internet	167
12.8	Summary	167

Appendix A	168

Index	195

Preface

Power system automation is at the cutting edge of technology in electrical engineering. It means having an intelligent, interactive power distribution and transmission network including:

- Increased performance and reliability of electrical protection
- Advanced disturbance and event recording capabilities aiding in detailed electrical fault analyses
- Display of real time substation automation in a central control center
- Remote switching and advanced supervisory control over the power network
- Increased integrity and safety of the electrical power network including advanced interlocking functions
- Advanced automation functions eg intelligent load shedding

At the conclusion of the reading of the book we hope that you will learn:

- The requirements for data communications in an electrical environment
- Suitability of different communication protocols for automation of power distibution and transmission networks
- New techniques in electrical protection, leading to increased reliability, performance and safety for personnel
- How to obtain extensive real-time information of your power network via the SCADA system, leading to informed decisions and productive use of manpower
- How to implement local and remote control of switchgear, including interlocking and intelligent load shedding
- How to effectively compare and critically analyze different products and systems available for protection, control and automation of electrical power networks

Typical people who will find this book useful include:

- Electrical engineers
- Control engineers
- Transmission and distribution engineers
- Design engineers
- Consulting engineers
- Power system engineers
- Protection engineers
- Technicians
- Maintenance supervisors

A basic knowledge of electrical principles is useful in understanding the concepts outlined in the book; but the contents are of a fundamental nature and are easy to comprehend.

The structure of the book is as follows.

Chapter 1: Introduction to power system automation. This chapter gives a brief overview of what power system automation means and the typical architecture of a system.

Chapter 2: Historical development of power system automation. The historical background from the perspective of the electrical protection, electronics, switchgear, and automation industries.

Chapter 3: Overview of power networks. An overview of power generation, transmission, distribution and network studies.

Chapter 4: Fundamentals of electrical protection. An overview of the electrical power systems protection world to place electrical network automation into context.

Chapter 5: Remote substation access and local intelligence. A discussion on the structure of the remote terminal unit (RTU), programmable logic controllers, protection relays and the intelligent electronic device (IED).

Chapter 6: Data communications. A review of data communications as it applies to electrical network automation with a discussion on topologies, communication techniques, media access control principles, the OSI Model and performance criteria.

Chapter 7: Communication protocols. A review of the typical protocols used such as DNP3, Modbus, Modbus Plus, Profibus, Ethernet, TCP/IP and the IEC 60870.5 standard.

Chapter 8: SCADA system. A discussion on the typical structure of a SCADA system.

Chapter 9: Communications in power system automation. The particular requirements for power system automation with a focus on protection, control, measurement and monitoring.

Chapter 10: Power system automation architectures. The different architectures (often referred to as type 1, 2, 3 or 4 systems) for power system automation are assessed here.

Chapter 11: Power system automation products on the market. The product offerings from GE, ABB, SEL, Siemens and Alstom are discussed here.

Chapter 12: Practical considerations. The various issues in using the new power system automation architectures and philosophies are considered here with a focus on basic strategies, suppliers and the impact of the Internet.

1

Introduction to power system automation

1.1 Definition of the term

Power system automation can be defined as a system for managing, controlling, and protecting an electrical power system. This is accomplished by obtaining real-time information from the system, having powerful local and remote control applications and advanced electrical protection. The core ingredients of a power system automation system are local intelligence, data communications and supervisory control, and monitoring.

(Note: Power system automation is also referred to as substation automation. The term 'substation' will be used throughout the text to describe mainly a building housing electrical switchgear, but it may also include switchgear housed in some sort of enclosure, for example a stand-alone ring main unit, etc.)

1.2 What is power system automation?

Power system automation may be best described by referring to Figure 1.1.

Power system automation, by definition, consists of the following main components:
- Electrical protection
- Control
- Measurement
- Monitoring
- Data communications

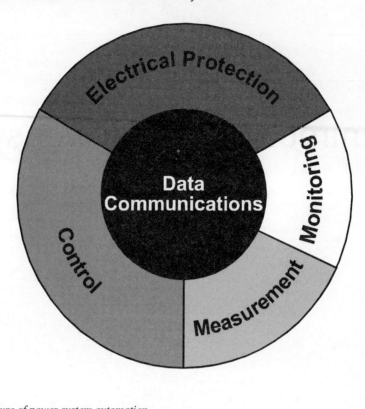

Figure 1.1
Functional structure of power system automation

1.2.1 Electrical protection

Electrical protection is still one of the most important components of any electrical switchgear panel, in order to protect the equipment and personnel, and to limit damage in case of an electrical fault.

Electrical protection is a local function, and should be able to function independently of the power system automation system if necessary, although it is an integral part of power system automation under normal conditions. The functions of electrical protection should never be compromised or restricted in any power system automation system.

1.2.2 Control

Control includes local and remote control. Local control consists of actions the control device can logically take by itself, for example bay interlocking, switching sequences, and synchronizing check. Human intervention is limited and the risk of human error is greatly reduced.

Local control should also continue to function even without the support of the rest of the power system automation system.

Remote control functions to control substations remotely from the SCADA (supervisory control and data acquisition) master(s). Commands can be given directly to the remote-controlled devices, for example open or close a circuit breaker. Relay settings can be changed via the system, and requests for certain information can be initiated from the SCADA station(s). This eliminates the need for personnel to go to the substation to perform switching operations, and switching actions can be performed much faster, which is a tremendous advantage in emergency situations.

A safer working environment is created for personnel, and huge production losses may be prevented. In addition, the operator or engineer at the SCADA terminal has a holistic overview of what is happening in the power network throughout the plant or factory, improving the quality of decision-making.

1.2.3 Measurement

A wealth of real-time information about a substation or switchgear panel is collected, which are typically displayed in a central control room and/or stored in a central database. Measurement consists of:

- Electrical measurements (including metering) – voltages, currents, power, power factor, harmonics, etc
- Other analog measurements, for example transformer and motor temperatures
- Disturbance recordings for fault analyses

This makes it unnecessary for personnel to go to a substation to collect information, again creating a safer work environment and cutting down on personnel workloads. The huge amount of real-time information collected can assist tremendously in doing network studies like load flow analyses, planning ahead and preventing major disturbances in the power network, causing huge production losses.

Note: The term 'measurement' is used in the electrical environment to refer to voltage, current and frequency, while 'metering' is used to refer to power, reactive power, and energy (kWh). The different terms originated because several different instruments were historically used for measurement and metering. Now the two functions are integrated in modern devices, with no real distinction between them; hence the terms 'measurement' and 'metering' are used interchangeably in the text. Accurate metering for billing purposes is still performed by dedicated instruments.

1.2.4 Monitoring

- Sequence-of-event recordings
- Status and condition monitoring, including maintenance information, relay settings, etc

This information can assist in fault analyses, determining what happened, when it happened, where it happened, and in what sequence (the place, time and sequence of a fault). This can be used effectively to improve the efficiency of the power system and the protection. Preventative maintenance procedures can be used by the condition monitoring information obtained.

1.2.5 Data communication

Data communication forms the core of any power system automation system, and is virtually the glue that holds the system together. Without communication, the functions of the electrical protection and local control will continue, and the local device may store some data, but power system automation system cannot function. The form of communication will depend on the architecture used, and the architecture may, in turn, depend on the form of communication chosen.

1.3 Power system automation architecture

Different architectures exist today to implement the components of power system automation. It is important to realize that not one single layout can exclusively illustrate a power system automation system. However, the most advanced systems today are developing more and more toward a common basic architecture. This architecture is illustrated in Figure 1.2.

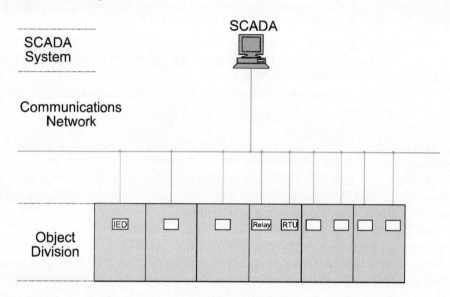

Figure 1.2
Basic architecture of power system automation

The modern system consists of three main divisions:

1.3.1 Object division

The object division of the power system automation system consists of intelligent electronic devices (IEDs), modern, third-generation microprocessor based relays and/or remote terminal units (RTUs). (PLCs also continue to play an important role in some systems). They receive analog inputs from the current transformers (CTs), voltage transformers (VTs) and transducers in the various switchgear panels, as well as digital inputs from auxiliary contacts, other field devices or IEDs, or the SCADA master. They can perform complex logical and mathematical calculations and provide an output either to the SCADA master, other field instruments or IEDs, or back to the switchgear to perform some command, for example open a circuit-breaker.

IEDs, modern relays, and RTUs are more comprehensively discussed in Chapter 5.

The object division consists of the process level (field information from CTs, VTs, etc) and the bay level (local intelligence in the form of IEDs, RTUs, etc). This is discussed further in Chapter 10.

1.3.2 The communications network

The communications network (comms network for short) is virtually the nervous system of power system automation. The comms network ensures that raw data, processed information, and commands are relayed quickly, effectively and error-free among the various field instruments, IEDs and the SCADA system. The physical medium will predominantly be fiber optic cables in modern networks, although some copper wiring will still exist between the various devices inside a substation.

The comms network needs to be an 'intelligent' subsystem in its own right to perform the functions required of it, and is not merely a network of fiber optic and copper wiring. Communications are discussed in more detail in Chapter 6.

The communication network serves as the interface between the bay level and the SCADA station level, which might be a SCADA master station in the substation itself, or remotely in a central control room.

1.3.3 SCADA master

The SCADA (supervisory control and data acquisition) master station(s) forms the virtual brain of the power system automation system. The SCADA master receives data and information from the field, decides what to do with it, stores it (directly or after some form of processing), and issues requests and/or commands to the remote devices. Therefore, the SCADA master is effectively in control of the complete power system automation system.

Now, a SCADA master consists simply of an advanced, reliable PC or workstation (with its peripheral and support hardware) and a SCADA software. (In contrast with a few years ago when SCADA systems used to run on big mainframe computers or some form of complex proprietary hardware.)

A SCADA master station may be installed in each substation of a power transmission network (station level), with all the substation SCADA stations forming part of a LAN or WAN (network level); or one SCADA master station may be directly in control of several substations, eliminating the station level.

SCADA systems are comprehensively covered in Chapter 8.

1.4 Summary

From the preliminary discussion in this chapter one can conclude that the principles of power system automation is not similar to process automation applied to a substation, but is in fact a totally different concept.

Chapter 2 reviews the historical development of power system automation, as it developed to its present day technological status due to, and sometimes parallel to, technological developments in various other related industries.

An overview of electrical power networks and the fundamental principles of electrical protection is necessary to fully appreciate the multitude of benefits a power system automation system can bring to the network, and to put it into perspective regarding electrical power transmission and distribution. These are the subjects of Chapters 3 and 4, respectively.

Chapter 5 is an introduction to the different devices used to gain remote access to substations through a SCADA system, namely RTUs, PLCs, and IEDs.

Data communication is central to any power system automation system. The principles of data communications are discussed in Chapter 6, and some of the most prominent and widely used communication protocols are reviewed in Chapter 7.

Chapter 8 briefly presents the requirements of the SCADA master station for power system automation.

Chapter 9 discusses the requirements of data communications in a power system automation environment. The main different architectures available today to implement a power system automation system are reviewed in Chapter 10.

Examples of some leading manufacturers' power system automation products are discussed in Chapter 11.

Practical considerations are looked at in Chapter 12, which concludes this text by summarizing the most important aspects to consider when looking at power system automation.

Appendix 1 discusses the Internet and some elements thereof, as the Internet has become a part of everyone's life, also in the workplace. Virtually every computerized society and industry is influenced by the Internet, and anybody involved in a technical field should be aware of the opportunities and restrictions of Internet technology. Power system automation also offers integration to the Internet, hence security considerations of this technology becomes critical.

2

Historical development of power system automation

2.1 Introduction

Power system automation, in its most advanced form today, has developed because of technological developments in a number …..of separate industries. The result is that different industries have noticed the potential of applying their new technologies to specific applications like substation switchgear, which contributed to the different architectures available today for power system automation. Herewith follows a short overview of the developments in these industries and their contribution to the concept of power system automation.

2.2 The electrical protection industry

Until very recently, the protection industry was developing the relays without any real cross-interfacing to the automation and communications industries specifically. This is because protection relays were always seen, inside and outside the protection industry, as a vital and integral part of any switchgear assembly, but as independent devices, without any need to communicate outside their zone of protection, much less outside the substation environment. However, this has changed dramatically over the last few years.

The first protection relays were the well-known (and still-used!) electromechanical relays. The most common ones were the IDMT (inverse definite minimum time) overload and earth-fault relays, which basically operated on the same electromagnetic induction principle, turning a metal disk to close a tripping contact when an overcurrent occurred. Thermal overcurrent relays were also used in later years, which relied on the heating properties of electric current to activate a tripping contact. The more 'sophisticated' relays of those early years were the current differential relays, which had to compare two current values to detect an electrical fault.

These early relays were quite stable and reliable, provided they were maintained regularly. Unfortunately, due to the large number of relays installed in most plants, maintenance was often neglected, with the result that reliability of the relays suffered.
(Note: Reliability = Trip when supposed to; Stability = Do not trip when not supposed to)
One of the more serious disadvantages of these relays was their lack of versatility, although it was not realized at the time. The characteristics of the relay were fixed, and to change these meant replacing the relay, often with major panel wiring modifications.

Another major disadvantage of the electromechanical relays, which was realized at the time, was their lack of accuracy. This often created major headaches for the protection engineer trying to achieve proper discrimination within tight boundaries. This lack of accuracy, together with the need to have more dependable, 'maintenance-friendly' relays, drove the research to develop more advanced relay designs.

Continuously improving designs, like installing RLC filters and using superior contact material like gold-alloys, served to increase the reliability, stability, and accuracy of the electromechanical relays. However, the first revolutionary breakthrough in relay design was the development of the electronic 'static' relay. (Named 'static' relays because they did not have any moving parts like the rotating disk of the electromechanical relay.) Although this new technology was not widely accepted, it marked a new direction in 'relay thinking', breaking away from the traditional, religiously accepted electromechanical standard.

The electronic static relay used the then new transistor-technology for relay operation. Measured electrical values such as currents and voltages were transformed to electronic values, which were then compared with the pre-set values to determine when the relay should trip. Thus the electronic relay was still an analog relay, as all functions were performed with analog values.

These relays offered greatly increased accuracy, reliability, and versatility, and were not nearly as maintenance-dependent as the electromechanical relays. They were mainly used in the transmission industry for more demanding and complicated applications, for example directional relays, voltage regulation, and differential relays. They did not gain wide acceptance in the distribution field to replace the conventional electromechanical overcurrent and earth-fault relays, mainly because of the following characteristics:

- These relays were introduced at a time when electronics was still regarded by many conservative engineers as a 'black art', best to stay away from.
- Their principles of operation were complex and difficult to understand for engineers and technicians who did not have an electronics background, and were therefore difficult to maintain and repair.
- They were expensive, and only became an economic consideration in complex applications.
- They were sensitive to power surges, glitches and voltage dips, which often required filters and surge protection to be installed, adding to the complexity and expense of the protection design.
- Their burden was relatively high, creating a big demand for auxiliary power, especially in large substations.

Protection manufacturers were still improving their designs and trying to achieve a wider acceptance of the electronic static relay when the second electronic revolution changed the world forever: The 'invention' of digital technology.

The first relays based on digital technology were soon to appear, with tremendous increases in versatility and reduction in costs. In addition, they used only a fraction of the auxiliary power the first electronic relays did, and due to economics these relays were designed with power filters, etc already incorporated, eliminating the earlier stability and reliability problems. These digital relays quickly gained wide acceptance because of their flexibility and affordability.

The protection settings of the first digital relays were done by dip switches on the front panel, and certain protection functions could be turned on or off simply by switching a dip switch, as well as changing the characteristic of the IDMT curve (normal, very or extreme inverse). They were virtually maintenance free, and due to their good mean-time-to-failure (except for a few manufacturers that experienced design problems), maintenance managers were willing to send the relays to the suppliers for repair, rather than training their own technicians and keeping spares to repair the relays in-house.

The first digital relays were limited in their processing power, with logical functions basically restricted to AND and OR gates (now termed first-generation digital relays). However, it did not take long for the first fully microprocessor-based relays to appear.

Microprocessor relays had previously undreamed of functionality. Accuracy, stability, and reliability became non-issues. Speed of operation became important with the advances in the switchgear market. Manufacturers now have started to equip their relays with metering and data storage capabilities. Now, with added programmability, came the need for communications capabilities, in order that the relay could be programmed via a laptop computer, and that data may be retrieved from it. The second-generation digital relays were developed, which were microprocessor-based, with limited communication capabilities.

Relay communications were initially intended to facilitate the installation, programming, and maintenance of the relays, with no thought of real-time communications at that stage.

The first communications to relays were very basic, with initially RS-232 and later RS-485 being used. Some manufacturers developed their own proprietary languages, for example SPABUS of ABB for their SPA series of relays. The electronic communication industry was also in its early stages of development at this stage, and the communication capabilities of relays tend to follow progress in the communications industry, which was, and still is, driven by demands from the process automation and computer industries.

Simultaneously, relays were being developed to be truly multi-functional, with more than one protection function (for example, IDMT overcurrent and earth-fault, directional and differential functions) being incorporated into one device; depending on manufacturer's different philosophies on the subject. Accurate metering capabilities, advanced communications, programmability and data storage capacity developed at a tremendous rate, to introduce the third generation of relays, which were multi-function relays with advanced communication capabilities. Finally, intelligent control functionality was incorporated, to foster what is known today as the intelligent electronic device (IED), as more fully discussed in Chapter 5.

New, more versatile and more powerful relays are continuously being developed today, limited not any more with technical possibilities, but rather by marketing demands, design philosophies, and the imagination.

2.3 The electronic industry

The 'first electronic revolution', namely the invention of the transistor, and the subsequent 'second electronic revolution', the invention of digital technology, were the cornerstones of literally all technological developments existing today. Televisions, personal computers, advanced motor-car engines, digital communications, space exploration, the list goes on and on: all made possible due to these two major technological breakthroughs.

Similarly, developments in the protection relay industry were possible due to these unique inventions in the electronic industry. Since then, developments in electronics have been the backbone and impetus of nearly all other technological advances in other industries, including the protection industry.

2.4 The switchgear industry

By nature, the switchgear and relays industries were always very closely associated, mostly existing as only different divisions of the same company, often with a very blurred separation between the two. Therefore, developments in these industries are closely related with one another.

Exciting developments took place in the switchgear market over the last few decades, with the development of vacuum and SF6 circuit-breakers, and more recently gas insulated switchgear (GIS) and internal arc proofing.

The impetus in switchgear development has been to increase ratings, reliability, speed of operation and safety, while at the same time decreasing cost and physical size. Protection relays in turn had to keep up with the demands from the switchgear divisions, especially concerning speed of operation, dependability, and reliability.

Switchgear became more sophisticated, and with the developments in the electronic industry, their control circuits became more and more advanced, eventually offering the possibility of interfacing with other electronic, and later digital, devices outside their immediate environment; giving birth to the concept of using RTUs and PLCs to gain remote access to switchgear operation.

2.5 The automation industry

The automation industry really came into existence with the release of digital technology to the world. Before that, automation was restricted to the old hardwired relay logic control panels, forming part of the electronic technician's area of expertise.

Digital technology soon fuelled the release of the first RTUs and SCADA systems, which later developed into large distributed control systems (DCS) for centralized plant control; as well as PLCs, which were specifically developed for localized process control. Originally, SCADA and PLC were seen as two opposing and competing concepts, with articles being published in abundance in technical magazines on the argument of 'SCADA vs PLC', and the advantages and disadvantages of each technology.

SCADA (and later DCS systems) were intended for centralized plant control, and hence were strong on supervisory control and communications to the remote units. This is where industrial communications really developed at a tremendous pace, and where the first communication protocols were introduced. The original function of the RTU was to collect information from the connected instruments in the field, translate the information into the language the supervisor required, send it through to the supervisor, receive instructions back, and relay this to the field instruments. Therefore, the RTU initially acted only as a communications buffer and translator between the supervisor and the field instrument.

Figure 2.1 illustrates a typical earlier SCADA network.

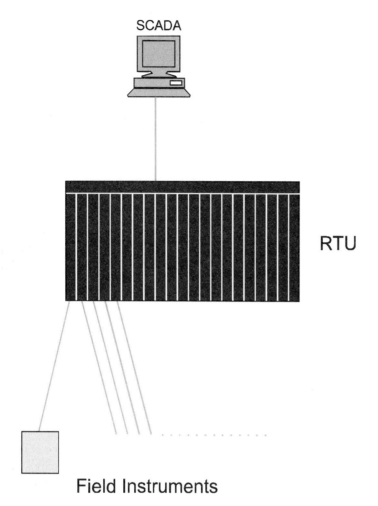

Figure 2.1
Typical early SCADA network

PLC technology was intended for localized process control, and hence was strong on local intelligence and well suited for processing analog signals (the well-known 0/4 –20 mA; ±10 V; etc standards). PLCs were originally intended to replace cumbersome relay logic panels, and were not intended to be exclusively controlled by a supervisor; therefore, they didn't have strong communication capabilities. PLCs functioned independently and the first 'supervisory' software mainly received information from the PLCs in order to keep the control room operator informed of what is happening in the plant (although this information was very limited, mostly restricted to current statuses, with historical data usually non-existent). Limited supervisory control was possible, and the first PLCs could not even be programmed remotely.

However, rapid technological development was at the order of the day, RTUs were bestowed with local intelligence, PLCs became more communication friendly, and the differences between the two technologies became smaller. The enlightened engineer realized that both these technologies have something to offer to his plant, and hybrid systems eventually were born, having both PLCs and RTUs communicating to a powerful supervisory system.

Figure 2.2 illustrates a typical control system in use today.

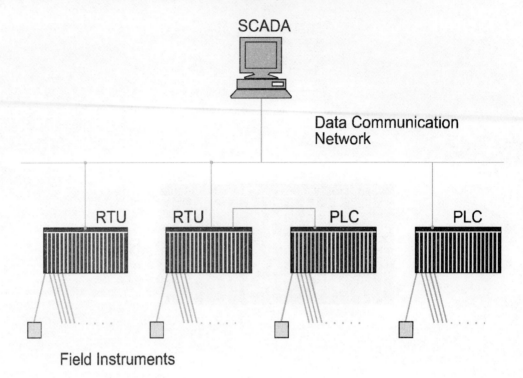

Figure 2.2
Typical control system in use today

A severe constraint until very recently upon using a combination of technologies was the fact that the first supervisory hardware and software, as well as the RTUs and PLCs, were exclusively proprietary, with every manufacturer developing his own communication protocol. Thus, products from different manufacturers could not communicate on the same network

Thankfully, another industry was progressing at an unforeseen and unbelievably rapid pace, namely the computer industry.

2.6 The computer industry

The digital computer is unarguably the most significant result of digital technology and probably the most important and influential invention in the history of humankind, although the tremendous impact digital computers would have on the world was not commonly foreseen at the time. The initial mainframe computers, although very powerful, were limited in their application, being mainly restricted to large corporations, universities, and government institutions because of the huge expense in obtaining, operating, and maintaining the equipment.

However, this changed with the introduction of the first personal computer. This step, making computer technology available and affordable to a large part of the world's population in their homes and at their workplaces, really sparked a virtual explosion of technological development. With independent computer workstations came the need for intercommunications and the first LANs (local area networks) and WANs (wide area networks) were established. The rest is all history.

Manufacturers of industrial automation products realized the benefits to them of making their SCADA software compatible with the normal PC, and soon PCs formed the hardware core of many SCADA systems, eliminating the need to have specific proprietary hardware. However, software was still very proprietary, with a specific manufacturer's software only able to talk to his own products, using a proprietary communications protocol.

Software developers quickly realized the enormous potential of SCADA software, which can communicate to various manufacturers' products and the first independent and open SCADA software packages were marketed, supporting various protocols. At last, this development made it possible to have one central supervisory control station, controlling equipment from different manufacturers.

Certain manufacturers, however, insisted on keeping their field equipment, SCADA software, and communications protocol proprietary, sometimes to their advantage, other times to their disadvantage. The question whether this approach would be more beneficial or detrimental in the long run, will soon become obsolete, as the drive toward an international communications protocol standard, is approaching finalization, and manufacturers whose equipment is not compatible to this standard, will soon disappear from the market. This will be discussed in more depth in Chapter 7.

2.7 The communications industry

Most of the major developments in the communications industry were fuelled by demands from other industries. For example, industrial communications developed due to a need for industrial automation, Ethernet developed due to the need for computer networking, other advanced protocols developed due to a need for high-speed communications in the motor manufacturing industry, etc.

Most of the time the speed of technological developments placed very high demands on the ingenuity and innovation of communication developers. Now, there is huge pressure on communications experts to increase the speed and bandwidth of the Internet. Therefore, the role the communications industry has played, often in the background, to make technological developments possible, should never be underestimated.

The most influential invention this industry has given the world is undoubtedly the fiber optic cable, which made possible the transfer of huge amounts of data over tremendous distances at the speed of light. Today's advanced power system automation systems cannot function without the use of fiber optics.

Similarly, satellite communications have played a key role in many technological developments, and still are a very underestimated and under-utilized communication medium today.

The power of radio communications has advanced tremendously over recent years, developing far past its initial function of voice communications, and very reliable, high-speed, high-capacity data communications are possible with modern radio links.

2.8 The measurement industry

The measurement industry has developed rather independently, utilizing and following the advances in other industries, rather than being a great influencing factor. However, this industry developed the first digital measuring techniques, converting analog values into digital format for the first digital recorders. These techniques are being used in modern relays today for metering purposes.

This industry also demanded the development of accurate current and voltage transformers for electrical metering purposes, which is integral to power system automation systems today.

3

Overview of power networks

3.1. Introduction

A typical electrical power network is illustrated in Figure 3.1.

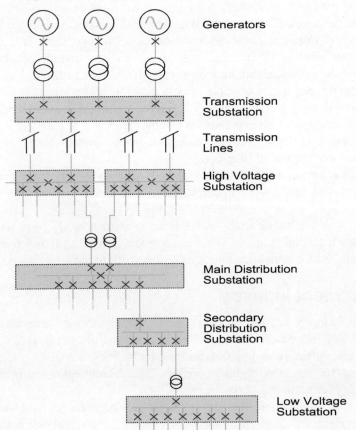

Figure 3.1
Typical electrical power network

An electrical network initiates at the point of generation. Electrical power is generated by converting the potential energy available in certain materials into electrical energy. Coal generators are still being used the most extensively worldwide, as well as oil and gas generators. However, these natural sources will not last forever, and other sources like wind, solar power, hydro-electricity, tidal energy, etc are being used more and more every day as sources of generating electrical energy. Nuclear power plants have tremendous potential, but a number of countries are against this form of generation, because of the inherent dangers in the process and the waste materials.

The electrical power generated is either transferred onto a bus to be distributed (small-scale), or to a power grid for transmission (larger scale). This is done either directly or through power transformers, depending on the generated voltage and the required voltage of the bus or power grid.

The next step is power transmission, whereby the generated electrical potential energy is transmitted via transmission lines, usually over long distances, to high-voltage substations. High-voltage substations will usually tap directly into the power grid, with two or more incoming supplies to improve reliability of supply to that substation's distribution network.

Figure 3.2 is a schematic illustration of a typical power grid.

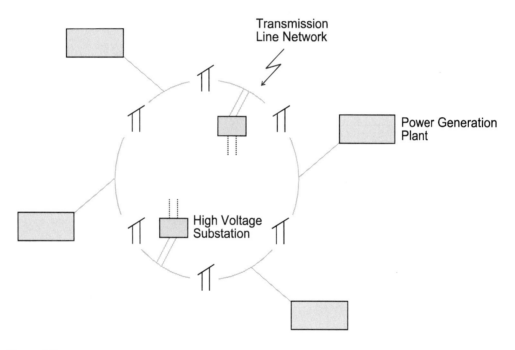

Figure 3.2
Typical power grid

Normally, the transmission voltage will be transformed at the high-voltage substation to a lower voltage for distribution purposes. Critical medium-voltage distribution substations will generally also have two or more incoming supplies from different high-voltage substations. Main distribution substations usually supply power to a clearly defined distribution network, for example a specific plant or factory, or for town/city reticulation purposes.

The distribution voltage is transformed to lower voltages again, either for lighting and small power applications, or for electrical motors, which is usually fed from a dedicated motor control center (MCC).

It can be seen from Figure 3.1 that the higher up in the hierarchy of a power network a substation is, the larger is the area that is affected by that substation. It is for this reason that utilities companies worldwide (who are responsible for the generation and transmission of electrical power) were the first companies that installed power system automation systems, as millions of dollars are lost in revenue in case of a substation blackout.

3.2 Power generation

Power plants usually consist of more than one generator simultaneously feeding into a power grid or network. Additional and sophisticated protection and control is needed, for example to ensure a generator is synchronized with the network before being switched on to the network, to ensure efficient loading of generators, to prevent 'power swings', to ensure network faults or overloading do not destabilize the generators, etc.

The current that flows in an AC (alternating current) generator depends on the generated voltage, on the phase angle of its internal voltage with respect to the phase angle of the internal voltage of every other machine in the system, and on the characteristics of the network and loads. Differences in the magnitude and/or phase angle of voltages between generators connected in parallel on the same network, will result in circulating currents flowing in the network, and one or more generators will run as a synchronous motor rather than a generator. This constantly changes the phase angles of the internal voltages with respect to each other, and results in instability of the network.

The issue of stability is the problem of maintaining the synchronous operation of the generators (and large synchronous motors) of the network. Disturbances on the system, caused by suddenly varying large loads, switching operations, occurrence of electrical faults, or loss of excitation in the field of a generator, may cause loss of synchronism. This is called transient instability.

Instability of a generator may also result from attempting to drive a generator beyond the limit of power the generator is capable of delivering; called the steady-state stability limit.

Owing to the complexity of generator protection and control, and especially control in large power plants, it is not surprising that electrical utility companies, were the first to use some form of power system automation, employing RTUs to facilitate switching operations, in addition to more sophisticated control systems and PLCs to control generator operation.

3.3 Power transmission

Once electrical power has been generated, it needs to be conveyed to locations where it is needed, usually over long distances. This is called power transmission. The higher the transmission voltage, the more economical is the transfer of electrical power. This can be seen from the well-known formula for electrical power, which states $P = \sqrt{3}\ VI \cos\phi$.

Therefore, the higher the voltage, the smaller the current needs to be for a certain power required, and the smaller the current, the smaller the required conductor size, hence the lower the cost of the transmission line. That is why transmission voltages tend to become higher and higher, as new technologies develop to achieve higher voltages, especially over very long distances. On the other hand, higher voltages mean more expensive equipment like transformers and switchgear, thus the most economical transmission network design will depend on a multitude of factors, like distance, available route, generated voltage, required end voltage, power demand, etc.

Overhead lines are far cheaper than underground cables for long distances, mainly because air is used as the insulation medium between phase conductors, and that no excavation work is required. The support masts of overhead lines are quite a significant portion of the costs, that is the reason why aluminum lines are often used instead of copper, as aluminum lines weigh less than copper, and are less expensive. However, copper has a higher current conducting capacity than aluminum per square millimeter, so once again the most economical line design will depend on a lot of factors.

Long lengths of transmission lines add significant inductive reactance to the network. This worsens the power factor ($\cos \phi$), as can be seen from the power triangle in Figure 3.3.

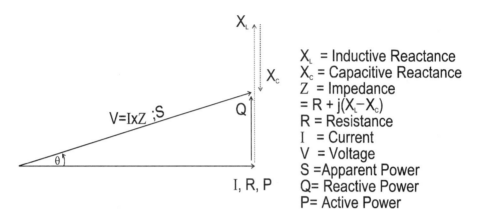

Figure 3.3
Power triangle

The lower the power factor (i.e. the larger the angle ϕ), the more inefficient the power transmission, and the higher the reactance losses. The installation of capacitor banks will improve the situation, as the added capacitance will oppose the buildup inductance (commonly called power factor correction).

The protection of transmission lines is more demanding than the protection of short feeder cables. This is because a fault may not always be very discernible to the protection scheme; due to the high resistance and reactance involved with long lines. Very sophisticated protection schemes have been developed in recent years, ranging from high-speed differential protection to medium-speed distance protection to low-speed sensitive earth fault protection.

Overhead lines are by nature prone to lightning strikes, causing a temporary surge on the line, usually causing flashover between phases or to ground. The line insulators are normally designed to relay the surge to ground, causing the least disruption and/or damage. This is of short duration, and as soon as it is cleared, normal operation may be resumed. This is why sophisticated autoreclosures are employed on an increasing number of overhead lines.

Power transmission is defined as the process of transferring generated power, normally at high voltages via overhead lines, to a high-voltage substation (generally, several high-voltage substations will form part of a transmission network). The voltage will then be transformed at the high-voltage substation to a lower value more suitable for distribution purposes, and from here on the process of transferring power is commonly referred to as power distribution.

Power transmission networks are normally quite complex, and efficient network control is a necessity.

3.4 Power distribution

At the high-voltage substation, the transmission voltage is transformed to a lower voltage for distribution purposes. This is because distribution is normally done over shorter distances via underground cables. The insulation properties of three-phase cables limit the voltage that can be utilized, and lower voltages, in the medium-voltage range, are more economical for shorter distances.

The high-voltage substation consists of the high-voltage switchgear and power transformers.

The substation will usually be fed from more than one supply and will feed to several main distribution substations, each main substation usually belonging to a specific client, and each main substation will normally be supplied from more than one high-voltage substation.

Main distribution substations then generally feed to a multitude of secondary substations. High-voltage and main distribution substations can be quite complex, with numerous incoming and outgoing feeders. In addition, protection schemes need to be sophisticated and reliable, to protect the expensive, critical switchgear, transformers, and feeders. This is an area where modern power system automation systems can have a multitude of benefits, offering the quality of protection needed, metering capabilities, data storage capacity, intelligent local control, and communication to a supervisory system.

The initial complex substation control systems were far too expensive to justify their use on distribution networks, due to the smaller area a distribution substation will affect in case of a power outage.

However, modern power system automation systems have changed this dramatically. Systems are available suitable, and justifiable, both for small distribution substations and large transmission substations.

3.5 Low voltage applications

Secondary medium-voltage substations typically feed through a number of power transformers to obtain the low voltage used in the factory. For example, in Europe 500 V is used for small to medium motors, and 380 V (220 V phase-neutral) for small power applications and lighting. In Australia, 433 V is used for small- to medium-sized motors, as well as for three-phase small power applications, with a resultant phase-neutral voltage of 250 V.

Motor control centers (MCCs) are becoming more intelligent in recent years, with more information being exchanged between the MCC and the PLC or DCS, than only start/stop signals as in the past, and with more local intelligence in the MCC itself.

3.6 Network studies

Network studies generally consist of load flow studies, fault level calculations, and stability studies.

A load flow study determines the voltage, current, power, and reactive power in various points and branches of the system under simulated conditions of normal operation. Load flow studies are essential in optimizing existing networks, ensuring an economical and efficient distribution of loads, and to plan future networks or additions to existing networks.

The currents that flow in different parts of a power system immediately after the occurrence of an electrical fault differ substantially from the current flowing in steady-state conditions. These currents determine the ratings of circuit breakers and other switchgear that are installed in the system, specifically the current flowing immediately after the fault and the current which the circuit breaker must interrupt. Fault calculations consist of predicting these currents for various types of faults at various locations in the system. The data obtained from fault calculations are also used to determine relay settings.

Stability studies predict the steady state and transient stability of the power system, taking the interaction of various generators and large synchronous motors on the network into account. Stability studies are very complex.

Network studies used to be extremely time-consuming in the pre-digital computer days, and often not very accurate. However, very powerful electrical analysis software in modern times has restricted human efforts to the input of equipment data and the interpretation of the study results. Network studies can now be done in a fraction of the time it used to take, with much more accurate results. This has greatly aided in designing, implementing, and maintaining more efficient power systems.

The real-time information that can be obtained from a power system automation system is a great aid to, and parallel with, network studies. Firstly, the steady-state information continuously obtained, archived and trended can tremendously assist the results of load flow studies. Secondly, the information obtained from disturbance records, as well as sequence-of-events recordings, can be used to determine the correctness of fault calculations and stability studies, as well as to verify the effectiveness of the electrical protection.

Therefore, the wealth of valuable information made available by a power system automation system can aid hugely in refining network studies and increase the predictability of the power network. This level of information was simply not available a few years ago.

4

Fundamentals of electrical protection

4.1 The need for electrical protection

It is not economically feasible to design and manufacture electrical equipment that will never fail in service. Equipment will and do fail, and the only way to limit further damage, and to restrict danger to human life, is to provide fast, reliable electrical protection. The protection of a power system detects abnormal conditions, localizes faults, and promptly removes the faulty equipment from service.

Electrical protection is not an exact science, but is rather a philosophy based on a number of principles. There are countless unique circumstances where protection is needed, and the techniques that will be applied have to take the specific conditions into account. Economic principles, namely the cost of the equipment that is protected, the cost of the protection equipment itself, the secondary cost of an electrical fault (for example, lost revenue or production losses), as well as probability analyses, all play a role in determining the protection philosophy that will be followed.

4.2 Overview of electrical faults

Electrical faults usually occur due to breakdown of the insulating media between live conductors or between a live conductor and earth. This breakdown may be caused by any one or more of several factors, for example, mechanical damage, overheating, voltage surges (caused by lightning or switching), ingress of a conducting medium, ionization of air, deterioration of the insulating media due to an unfriendly environment or old age, or misuse of equipment.

Fault currents release an enormous amount of thermal energy, and if not cleared quickly, may cause fire hazards, extensive damage to equipment and risk to human life. Faults are classified into two major groups: symmetrical and unbalanced (asymmetrical). Symmetrical faults involve all three phases and cause extremely severe fault currents and system disturbances. Unbalanced faults include phase-to-phase, phase-to-ground, and phase-to-phase-to-ground faults. They are not as severe as symmetrical faults because not all three phases are involved. The least severe fault condition is a single phase-to-ground fault with the transformer neutral earthed through a resistor or reactor. However, if not cleared quickly, unbalanced faults will usually develop into symmetrical faults.

Switchgear need to be rated to withstand and break the worst possible fault current, which is a solid three-phase short-circuit close to the switchgear. 'Solid' means no arc resistance. Normally arc resistance will be present, but this value is unpredictable, as it will depend on where exactly the fault occurs, the actual arcing distance, the properties of the insulating medium at that exact instance (which will be changing all the time due to the heating effect of the arc), etc. Therefore, in fault calculations, the arc resistance is ignored, as it is undeterminable, with the result that the worst case is calculated. (The arc resistance will tend to decrease the fault current.)

4.3 Protection components

4.3.1 Fuses

Probably the oldest, simplest, cheapest, and most-often used type of protection device is the fuse. The operation of a fuse is very straightforward: The thermal energy of the excessive current causes the fuse-element to melt and the current path is interrupted. Technological developments have made fuses more predictable, faster, and safer (not to explode).

A common misconception about a fuse is that it will blow as soon as the current exceeds its rated value (i.e. the value stamped on the cartridge). This is far from the truth. A fuse has a typical inverse time–current characteristic as illustrated in Figure 4.1. The pick-up value only starts at approximately twice the rated value, and the higher the current, the faster the fuse will blow.

By nature, fuses can only detect faults associated with excess current. Therefore, a fuse will only blow in earth fault conditions once the current in the faulty phase has increased beyond the overcurrent value. Therefore, fuses do not offer adequate earth fault protection. A fuse has only a single time–current characteristic, and cannot be adjusted. In addition, fuses need to be replaced after every operation. Finally, fuses cannot be given an external command to trip.

Fuses are very inexpensive. Therefore, they are suitable to use on less critical circuits and as backup protection should the main protection fail, offering very reliable current-limiting features by nature.

Another advantage of fuses is that they can operate totally independently, that is, they do not need a relay with instrument transformers to tell them when to blow. This makes them especially suitable in applications like remote ring main units, etc.

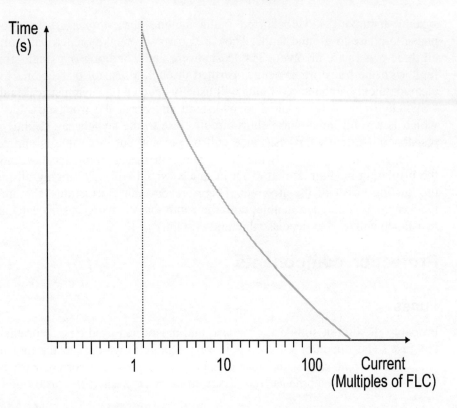

Figure 4.1
Typical time–current of a fuse

4.3.2 The relay/circuit-breaker combination

The most versatile and sophisticated type of protection available today, is undoubtedly the relay/circuit-breaker combination. The relay receives information regarding the network mainly from the instrument transformers (voltage and current transformers), detects an abnormal condition by comparing this information to pre-set values, and gives a tripping command to the circuit-breaker when such an abnormal condition has been detected. The relay may also be operated by an external tripping signal, either from other instruments, from a SCADA master, or by human intervention.

The most reliable way to provide auxiliary power to the relay is by way of a battery-tripping unit (BTU). The unit basically consists of a set of batteries, which supplies DC power to the relay and trip circuit. The batteries are kept under charge by a battery charger, which normally is rated to have enough capacity to supply the standing load of the switchgear panel (relay auxiliary power, indication lamps, etc). When temporary high current is needed, usually to provide a circuit-breaker tripping supply, the batteries will supply this, and be recharged after the event. The batteries then also function as a full backup in case of total power failure.

The AC input to the BTU is usually supplied from the panel VT, or from a lighting transformer.

Sometimes the total power requirements of the switchgear panel will be supplied from the BTU, and sometimes only the tripping circuits and relay. The reasoning behind the latter designs is to keep the BTU as small and inexpensive as possible, and that the closing supply to circuit breakers, as well as indication lamps, etc, will not be used in case of a power failure anyway. This really depends on the electrical operating philosophy regarding the specific substation and the design philosophy of the electrical engineer. There are pros and cons both ways, and it is not the purpose of this text to examine those in detail. It will only be mentioned that to supply DC power with battery backup to all the switchgear panel auxiliaries, is an advantage for maintenance purposes.

However, what often occurs in practice, and which is very bad engineering practice, is to power the relay and trip circuit directly from the panel VT. This will function correctly in most instances, but when a really severe three-phase short-circuit occurs, the voltage of the substation may drop dramatically, causing malfunction of the tripping circuit.

There are other ways to overcome this, like capacitive tripping circuits, and AC series tripping schemes, but each with its own disadvantages, and none as reliable as the DC shunt tripping arrangement.

BTUs have become very advanced in the electronic era, with current limiting, voltage regulating, very smooth changeover to battery power, advanced monitoring alarm functionality, etc.

The circuit-breaker opens it main contacts when the tripping signal is received, interrupting the current. Initially, circuit-breakers used air as the insulating medium, later insulating oil (the oil also acting as a cooling medium), and now vacuum or SF6 (sulfur-hexafluoride) gas. Modern switchgear is smaller with higher current-interrupting capabilities than in earlier years.

Gas insulated switchgear (GIS) is gaining in popularity, especially in countries where space is a limitation. GIS means that not only the moving contacts, but all the current-carrying parts, including the busbars, are enclosed in SF6-gas.

The connection between the relay and the circuit-breaker trip coil is purely electrical. This used to be one possible weak link in the trip circuit. One popular method to increase the reliability of the trip circuit for critical substations is to provide a full back-up trip supply. A backup trip coil is installed in the circuit-breaker, with backup protection, powered by a second, independent BTU.

Often a second set of contacts of the same relay is used for the backup protection, which defeats the purpose of having full redundancy somewhat, as the relay itself then forms the weakest link.

A second method of increasing the reliability of the trip circuit is to incorporate a trip circuit supervision relay, which continually monitors the continuity of the trip circuit, and activates an alarm when an unhealthy trip circuit is detected. One shortcoming of this method is that it cannot monitor the main protection relay itself.

Modern relays now have advanced self-monitoring and trip circuit supervision functionality, activating an alarm when it detects a fault within itself or in the trip circuit, increasing reliability of the complete trip circuit tremendously. These new-generation relays ensure that the question of reliability is not the major headache it used to be to the protection engineer, and more advanced questions now demand his attention.

The functionalities of modern relays are examined more closely in Chapter 5, as this forms a crucial cornerstone of the concept of power system automation.

4.3.3 Instrument transformers

Relays need information from the power network in order to detect an abnormal condition. This information is obtained via voltage and current transformers (collectively called instrument transformers), as the normal system voltages and currents are too high for the relays to handle directly, and the instrument transformers protect the relay from system 'spikes' to a certain extent.

Voltage transformers (VTs) are often used additionally to supply AC voltage to the substation BTU, circuit-breaker closing coils, indication lamps, alarms, etc. Design of voltage transformers is very specialized, and certain VT designs are not suitable for certain applications, for example where the load on the VT is going to be very low, etc.

Therefore, the electrical/protection engineer should ensure that he or she specify the system conditions and application of the VT as correctly and detailed as possible.

Two types of voltage transformers are commonly used: the electromagnetic type (most common), which operates on the same principle as the power transformer, and the capacitive type (less common), which uses the formula $V_s = V_p \times C_2/(C_1 + C_2)$, as illustrated in Figure 4.2.

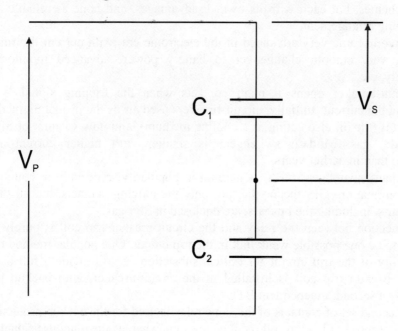

Figure 4.2
Capacitive voltage transducer

Current transformers (CTs) are classified into two main groups, according to their application: Metering CTs and protection CTs. The very important difference is as follows (refer to Figure 4.3):

- Metering CTs are designed and manufactured to be very accurate (usually within 0.5%) from very low currents up to approximately 140% of full load currents (depending on which international standard is used). The CT will saturate at higher currents, meaning that the secondary current will stay constant, irrespective of the primary current, for two reasons. Firstly, it is much more expensive to manufacture CTs that have a linear curve from very low to very high currents, and secondly, the saturation of the CT protects the sensitive metering equipment connected to it from high currents.

Fundamentals of electrical protection 25

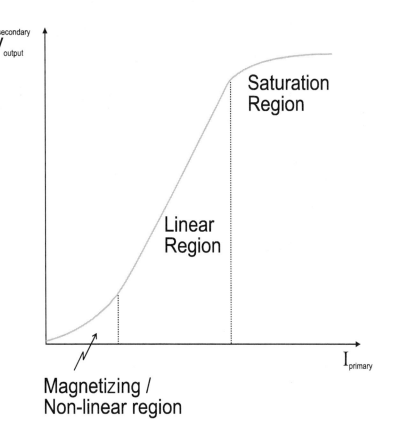

Figure 4.3
Typical CT magnetizing curve

- Therefore, the linear region of a metering CT should reach from as low current value as possible to 140% of full load current (FLC). The CT should saturate at higher primary currents to protect the sensitive metering equipment.
- Metering CTs are usually connected in series in the primary circuit for low currents, as this method is more accurate. However, for high current applications, the ring type CT (or bar type for busbars), are used, as no current then passes directly through the CT.
- Protection CTs are designed to saturate at very high levels, in order that the secondary current stays accurate in the fault current range to provide correct information to the protection relays. The linear region of the CT should therefore extend to approximately 12 × FLC, depending on the specific protection application. CTs for differential protection applications need to be more accurate as for normal overcurrent applications, and in addition need to be matched.

However, protection CTs are, by nature of their design, not as accurate as metering CTs, especially in the very low current range, where they tend to be very inaccurate because of the non-linearity of the curve in this region (refer to Figure 4.3). Another reason for this traditional inaccuracy of protection CTs is the high burden that analog protection relays placed on the CT. This burden becomes a significant factor with low0. primary currents, leading to severe inaccuracies. Additionally, the burden varies with the tap setting, making it very difficult to compensate for the resultant metering inaccuracy.

Modern relays are performing metering functions as well, although they are not intended (yet) to replace accurate revenue metering. However, they are used more and more for accurate metering information, and as they use only one set of CTs for metering and protection, these CTs are required more and more to be both accurate for low currents and linear for fault currents.

Another advantage of digital relays is that their burden on the CTs is small. This makes it much easier to use one set of CTs for protection and accurate metering. This is indeed becoming increasingly the case when using modern, intelligent digital relays, and manufacturers of these relays guarantee accuracy of up to 0.1%, not including CT inaccuracy.

4.4 Protection qualities

4.4.1 Overview

The basic function of electrical protection is to detect system faults and to clear them as soon as possible. For any one particular application, there are many ways to do this function, with varying degrees of effectiveness. Some schemes will be expensive and of excellent technical performance, while others will be less costly, but not as adequate. The choice is influenced by the overall protection philosophy of the plant, and the importance of the equipment or portion of the network to be protected, weighing cost against performance.

The general philosophy of applying protection in a power network is to divide the network into protective zones, such that the power system can be adequately protected with the minimum part of the network being disconnected during fault conditions. The zones can either be very clearly defined, with the protection operating exclusively for that zone only (as in differential protection, illustrated in Figure 4.4), or less clearly defined, with overlapping of the protection function between zones (for example, overcurrent protection, as illustrated in Figure 4.5).

When protective zones overlap, the main protection for one zone usually functions as the backup protection for the lower zone. (For critical applications, a specific zone will have its own backup protection, either an exact duplication of the main protection, or having, for instance, overcurrent protection as a backup for differential protection.)

Fundamentals of electrical protection 27

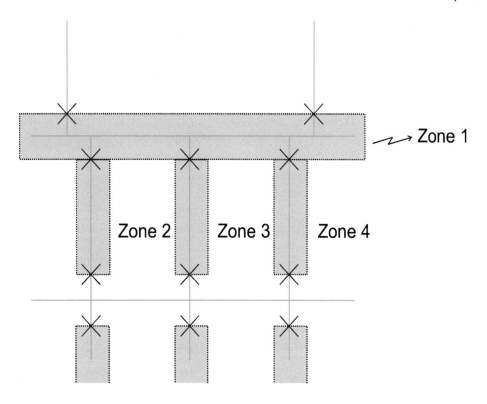

Figure 4.4
Clearly defined zones

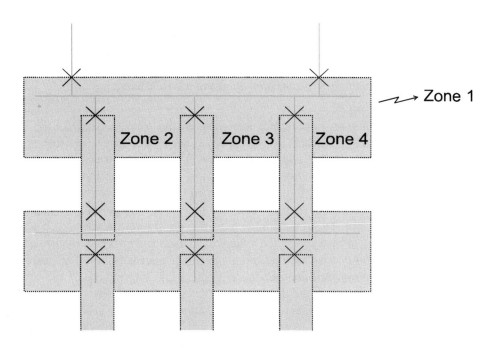

Figure 4.5
Overlapping zones

4.4.2 Discrimination

Discrimination, or selectivity, is the ability of the protection to isolate only the faulted part of the system, minimizing the impact of the fault on the power network.

Absolute discrimination is only obtained when the protection operates exclusively within a clearly defined zone. This type of protection is known as 'unit protection', as only one unit is exclusively protected for example, a transformer, or a specific feeder cable. The term 'zone protection' is also commonly used.

Unit protection can only be achieved when the following essentials are satisfied:
- Sensing or measuring devices must be installed at each (electrical) end of the protected equipment; and
- There has to be a means of communication between the devices at each end, in order to compare electrical conditions and detect a fault when present.

The most common form of unit protection is current differential protection, whereby current values at each end of the protected equipment is measured and compared, and a trip signal is issued when the difference in measured values is more than a predefined threshold value. Differential protection is discussed a little more in section 4.5.3.

(Note: Differential protection operates according to Kirchoff's law, which states that 'current in' must equal 'current out', compensating for a small percentages of losses due to leakage reactance.)

The main advantages of unit protection are:
- Only the faulted equipment or part of the network is disconnected, with minimum disruption to the power network.
- Unit protection operates very fast, limiting damages to equipment and danger to human life. Fast operation is possible because the presence or absence of a fault is a very clear-cut case.
- Unit protection is very stable (see next section).
- Unit protection is very reliable (provided the communication path is intact).
- Unit protection is very sensitive.

The major disadvantages of unit protection are the following:
- It is very expensive.
- It relies on communication between the relays installed at either end. Unit protection for cables, for instance, need pilot wires (or fiber optic cables) to run the whole length of the cable. These pilot wires can easily be, and often are, damaged, causing malfunction of the protection scheme. Some earlier relays were configured to trip when the communication to the slave relay was lost, resulting in poor stability (see next section for a brief discussion of stability). The reasoning behind this configuration was that it was seen as the lesser of two evils, the other 'evil' not to trip when required to. Therefore, reliability was seen as more important than stability. The correctness of this reasoning will depend on the application, importance of the equipment to be protected, and the detrimental effect of poor stability vs the detrimental effect of poor reliability. Most of the later, more advanced relays had pilot wire supervision incorporated, giving early warning of a communication problem. Modern relays have communication supervision and self-diagnostic functions, overcoming these earlier problems.

- It can be maintenance-intensive to keep the communication medium intact, depending on the application and environment.

The discrimination qualities of non-unit protection are not absolute, as the relay functions independently and will generally operate whenever it sees a fault, no matter where the fault is located.

Therefore, to achieve proper discrimination for non-unit protection schemes, the principle of grading is applied. Consider the example, as illustrated in Figure 4.6, where the protection consists of only overcurrent relays. (This only serves as a theoretical example to demonstrate the principle of grading. The zones will overlap, but for clarity purposes have been shown as discrete.)

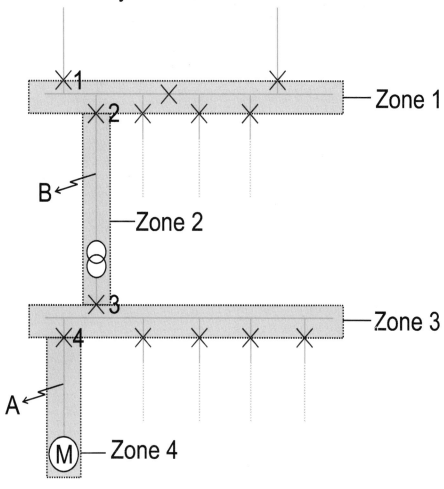

Figure 4.6
Protection example

The main functions of the relay in the diagram are as follows:
Relay 4: Protect the feeder cable to the motor and the motor itself (zone 4)
[Specialized motor protection is ignored for this example]
Relay 3: Protect the low voltage switchgear panel and busbars (zone 3)

Relay 2: Protect the feeder cables and transformer (zone 2)
Relay 1: Protect the high voltage switchgear panel and busbars (zone 1)

Modern digital relays still primarily simulate the inverse time–current characteristic of the electro-mechanical relay, for three main reasons:

- Firstly, this characteristic has proven to be very effective for most protection applications.
- Secondly, protection engineers and technicians are very familiar with this characteristic.
- Thirdly, and the most important, new relays have to grade with older relays still in service, and the inverse time–current characteristic has become an international standard for protection relays world-wide.

If the relays in our example were all of the same type, and no lower or upper restrictions were placed on the grading, it would be quite simple, and the time–current would look something like the graph in Figure 4.7 (assume a normal inverse characteristic with high-set):

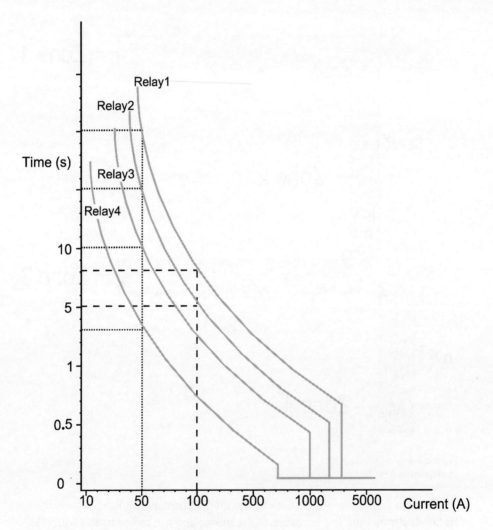

Figure 4.7
Overcurrent grading example

Normally, a factor of 0.5 should be applied between adjacent curves to ensure proper discrimination. This is necessary to allow for inaccuracies in instrument transformers and relay operation, circuit-breaker operating times, as well as uncertainties in network and equipment reaction under fault conditions.

This is easily achieved in Figure 4.7, as follows (the time settings have been exaggerated for illustrative purposes):

- Fault at point A, causing a fault current of 50 A (referred to primary side) to flow: Relay 4 will trip in 4 s, clearing the fault, without any of the up-stream relays tripping. Should relay 4 (or the circuit-breaker) fail to trip, relay 3 will trip in 10 s, clearing the fault. (This is the principle of backup protection.)
- Fault at point B, causing a fault current of 100 A to flow: Relay 2 will trip in 5 s, clearing the fault. Relay 3 and 4 will not see the fault, and hence will not trip. Should relay 2 (or the circuit-breaker) fail to trip, relay 1 will trip in 8 s, clearing the fault.

Therefore, proper discrimination is obtained in this example.

However, the following restrictions will in practice be present, even for a simple network like this:

- An upper limit will be imposed on relay 1 by the protection settings of the utility company
- A lower limit for relay 4 exists in order not to cause tripping of the overcurrent relay with motor startup
- An upper limit for relay 2 exists to avoid permanent damage to the transformer

Imposing these limitations on the protection graph will complicate the design quite dramatically, as illustrated in Figure 4.8:

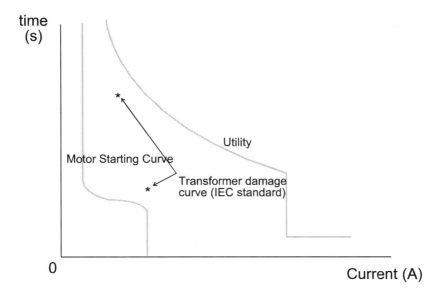

Figure 4.8
Imposing time–current restrictions

It can be seen that it will not be practically possible to fit the four relay graphs in between the lower limit of the motor starting curve and the upper limit of the utility's protection curve, with relay 2's curve clearing the transformer damage curve, and still achieve proper discrimination.

Situations like this necessitate the use of other types of protection, for example use an overcurrent relay with a very inverse characteristic for relay 4, replace relay 3 with a bus-zone protection scheme ('special' application of differential protection, see section 4.6.5.); replace relay 2 with a current differential scheme, and replace 1 with a bus-zone protection scheme.

It will be seen that there are many different ways to overcome the discrimination problem as per the example, and that is why there are no definite, clear-cut rules, but only good engineering principles and guidelines to apply to the specific circumstances.

Digital relays offer increased accuracy and flexibility, greatly assisting in overcoming discrimination problems. A third-generation digital relay will offer the possibility of customizing a curve to overcome the restrictions as in Figure 4.8.

4.4.3 Stability

Stability, also called security, is the ability of the protection to remain inoperative for normal load conditions (including normal transients like motor starting).

Most stability problems arise from incorrect application of relays and lack of maintenance.

4.4.4 Reliability

Reliability, or dependability, is the ability of the protection to operate correctly in case of a fault.

Reliability is probably the most important quality of a protection system.

4.4.5 Speed of operation

The longer the fault current is allowed to flow, the greater the damage to equipment and the higher the risk to personnel. Therefore, protection equipment has to operate as fast as possible, without compromising on stability. The best way to achieve this is by applying unit protection schemes. However, unit protection is expensive, hence the importance and cost of the equipment to be protected, and the consequences of an electrical fault, must be considered and weighed against the cost of very fast protection schemes.

4.4.6 Sensitivity

The term sensitivity refers to the magnitude of fault current at which protection operation occurs. A protection relay is said to be sensitive when the primary operating current is very low. Therefore, the term sensitivity is normally used in the context of electrical protection for expensive electronic equipment, or sensitive earth leakage equipment.

4.5 Main functions of protection relays

The following is not meant to be an exhaustive list of protection relay functionality, but only briefly discuss some of the most commonly used functions, in order to illustrate the advantages and increased functionality of digital relays.

4.5.1 Overcurrent

The overcurrent relay typically displays the inverse definite minimum time (IDMT) characteristic as displayed in Figure 4.9. Traditionally, normally inverse (NI), very inverse (VI), and extremely inverse (EI) have been applied, with each type of curve characteristic to a specific type of relay. Multitudes of curves, up to 15 in one relay, user selectable, are available with modern relays.

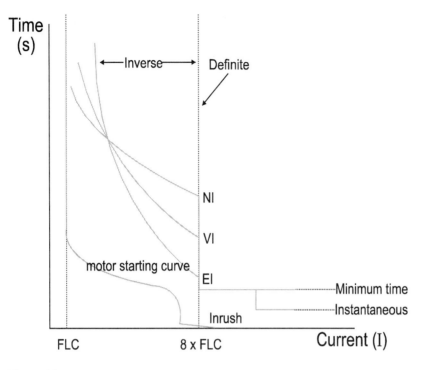

Figure 4.9
Traditional overcurrent curves

The overcurrent function mainly covers two eventualities:

Overload
This is not strictly a fault condition, but occurs when equipment is electrically overloaded in 'normal' system conditions. This may happen due to non-electrical problems, for example a bearing failure on a motor, or due to an inefficient electrical design. Whatever the cause, the overloaded equipment must be disconnected from the network, to protect both the equipment and the network, and to provide secondary problems, for example fire hazards.

Overload protection can be provided on the inverse part of the time–current characteristic. However, this curve characteristic does not reflect the temperature rise in the motor closely enough, and for this reason a thermal overload function is normally used for overload protection. An extremely inverse characteristic will come close to approximate the thermal characteristic of the motor, and the higher the overload current, the sooner it will be disconnected, which is the strength of the inverse time–current characteristic.

Thermal overload elements are very suitable for relatively low but sustained overloading.

Fault current
The overcurrent relay will pick up a phase-to-phase fault and a three-phase fault. Any phase-to-ground fault will be picked up eventually if not cleared by earth fault protection. This type of fault normally results in very high currents flowing within a few cycles. This

is the specific function of the high-set element of the overcurrent relay. The high-set element is usually set for between four times and eight times of normal full load current, depending on the load characteristics of that specific feeder (allow for starting of large motors, etc). The minimum tripping times achievable with electromechanical relays used to be 0.1 s. Digital relays generally offer a minimum setting of 0.01 s. This is 10 ms, which means that the short circuit can be detected within the first cycle. However, the fault will not be cleared in such a short time, as the circuit-breaker operating time still adds to the equation (minimum of 60 ms for most modern high-speed circuit-breakers).

Digital relays achieve this fast detection of fault current by comparing the rate of current rise during the current waveform.

The high-set function is also commonly called 'instantaneous overcurrent protection'.

4.5.2 Earth fault

Phase-to-earth faults are covered by earth fault relays. The most common form of earth fault protection operates on the principle that the vector sum of currents flowing in a balanced three-phase system equals zero. A very effective combination of overcurrent and earth fault protection has developed in the era of electromechanical relays, and the same principle is still used today in most protection schemes. This is illustrated in Figure 4.10. Only two phases need to be monitored by the overcurrent relay, the reason being that a fault on the third phase will be either to one of the other two phases, or to earth. A phase-to-earth fault will cause an unbalance in the three phases, resulting in a current flowing in the earth fault element, tripping the earth fault relay. The same protection CTs are thus being used in this arrangement.

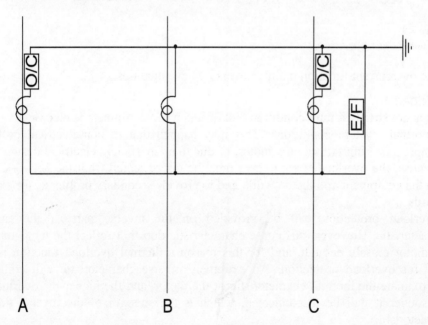

Figure 4.10
Economical CT arrangement for O/C and E/F protection

Earth fault protection is by nature more sensitive than overcurrent protection, and will clear a fault a lot quicker. That is the reason, for example, that certain types of power cable are individually screened, with the metal screen earthed during installation. This ensures that a cable fault will be a single phase-to-earth fault before a three-phase fault develops, in order that the fault can be cleared quickly by the earth fault protection.

It is very good engineering practice to have earthed phase segregation as far as possible in all equipment to achieve the same result; however, this means substantial added costs, and again comes down to the question of the 'economical' protection philosophy.

Owing to natural system unbalances (leakage reactance, inrush currents to transformers and motors, etc) and CT and relay inaccuracies, the earth fault setting can seldom be lower than 10–20% of full load current, depending on system conditions, otherwise stability of the relay is compromised, and 'nuisance tripping' may occur.

A more specialized type of earth fault protection commonly used, is 'restricted earth fault' protection, where the current flowing from earth into the star-point of a neutral earthed transformer is monitored. Ideally this current should be zero under normal system conditions, with a current flowing only under earth fault conditions. This is a more accurate form of earth fault protection, but can by nature only be applied in the transformer zone of protection.

Another form of earth fault is often used in especially transmission applications, termed 'sensitive earth fault' or 'sustained earth fault' protection. This is to provide for very low earth fault currents flowing for a sustained period of time, for example an overhead line conductor lying on dry soil. The setting is very low, normally in the range of 2–8 A, with a relatively long time delay in the range of 4–6 sec.

(Note that this text does not cover low voltage reticulation applications, and hence a discussion on sensitive earth leakage protection to protect personnel against direct electrical contact is not relevant.)

4.5.3 Differential protection

Differential protection schemes vary according to the type of equipment to be protected, the most common being machine and feeder differential protection. The protection relays differ in their compensation methods for typical internal losses in the equipment to be protected, but operates on basically the same principle. The values of current going into and out of the equipment are measured and compared. The relay trips if the difference in current exceeds a pre-set value, compensating for internal losses in the equipment and CT inaccuracies. Figures 4.11(a) and (b) illustrate the use of a differential protection scheme.

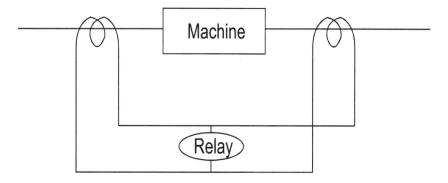

Figure 4.11(a)
Machine differential protection

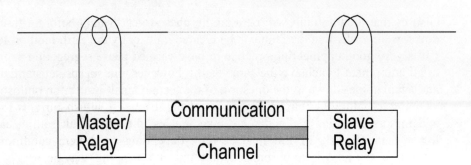

Figure 4.11(b)
Feeder differential protection

With machine differential protection (motors or transformers), the sets of CTs are close to each other, and only relay needs to be used in most cases, with current flowing through the relay in case of a difference in current values. The CTs need to be matched as closely as possible, and in most cases a biased scheme will be used, where the relay will compensate for the difference in measured current due to CT inaccuracies and other losses. It will be noted that compensating for these differences is much easier achieved with programmable digital relays than it used to be with electromechanical relays.

With feeder protection, the two sets of CTs are far away from each other. Two relays are installed, one at both ends of the equipment, one master and one slave. The slave relay measures the current at its end and sends it through to the master relay via the communication channel. The master relay compares the values and decides whether a trip should be executed. The master relay sends a trip signal to the circuit-breaker and, depending on the configuration, also sends a tripping command to the slave relay to trip the circuit-breaker at the other end.

This protection scheme is utterly dependent on the integrity of the communication channel between the relays. This channel used to be copper pilot wires, sometimes as pilot wires incorporated in power cables, and sometimes as a separate signal cable. Although copper pilot wires are still extensively used, especially for short distances like transformer differential protection, fiber optic cables are being used more and more with the additional capabilities of digital relays, especially for long distances and high-speed applications.

It is important to note that the communication link between the master and slave relay needs to be a direct one-on-one link, that is, not through a network or a SCADA system, as speed of communication (in ms) and absolute integrity of the link are vital to the reliable operation of differential protection and can not be compromised.

Table 4.1 summarizes and compares the qualities of the different types of direct communication links.

Busbar zone protection is a particular more complex form of current differential protection, as more than two values need to be taken into account, and is discussed in section 4.6.5, along with other forms of busbar protection.

Properties	Independent Copper Cable	Pilot Wires in Power Cable	Fiber Optic Cable
1. Integrity	Very high	Very high	High (Lower than copper cable due to equipment needed)
2. Cost	Low, but proportionally higher with longer distances due to resistance (larger sizes needed)	Medium to high, depending on configuration and distance	High for low distances, economical for long distances 1
3. Susceptibility to mechanical damage	Low to medium, depending on installation	Low 2	Medium to high, depending on installation
4. Susceptibility to electromagnetic interference	Medium, depending on distance from power cables 3	High, can be improved with screening, but added costs	None
5. Susceptibility to lightning	High	Medium	None
6. Nature of maintenance	Easy	Very Difficult	Specialized
7. Recommended use	Short distances, 'in-house'	High risk of mechanical damage present, or severe space restrictions	Long distances, outside use in lightning risk areas
8. Recommended installation	Protect against mechanical damage; Clearance from power cables, especially medium and high voltage cables	Not applicable	Protect against mechanical damage, use experienced personnel/contractor

Table 4.1
Communication links for differential protection

Notes:

Plastic fiber may be used as an economical option for distances up to 300 m, but not suitable for longer distances due to higher losses compared to glass fiber.

Damage to the pilot wires inside a power cable will probably be accompanied by damage to the power cable as well.

4.5.4 Voltage regulation

Sophisticated voltage regulation protection is mainly used in power generation, mainly to protect the grid from over-voltage or under-voltage from a specific generator; otherwise the stability of the grid may be compromised.

Under-voltage protection is often used in distribution applications, to prevent specific undesirable 'power-up' occurrences after a power blackout.

Under-voltage protection is also extensively used in motor protection applications; and over-voltage protection to protect sensitive electronic equipment for instance.

4.5.5 Frequency regulation

Frequency regulation, that is, overfrequency and underfrequency protection is mainly used in generation applications.

Underfrequency relays are often used in automatic load-shedding applications, as a drop in frequency is an early sign that a power network is being overloaded.

4.5.6 Distance protection

Distance protection is mainly used in transmission applications as an alternative, and often complementary, to differential protection, especially for very long lines.

Distance protection basically operates on the principle that the impedance of a line is stable under healthy line conditions. An electrical fault on the line will cause the impedance to alter. This will result in a shift of the voltage and current phasors with respect to one another. This shift will be detected by the relay.

Digital relays offer much increased accuracy, stability, and reliability for distance protection schemes.

4.5.7 Negative sequence protection

A healthy power system has balanced three-phase currents flowing, called positive sequence currents. When a disturbance occurs on the system, the fault currents can be shown by complex mathematical calculations to consist of a combination of positive sequence, negative sequence, and zero sequence currents.

The microprocessor power of digital relays has enabled them to perform the breakdown of measured currents accurately in their positive, negative and zero sequence components. Two-phase and three-phase faults all involve negative sequence currents. Thus the detection of negative sequence currents offers much faster and more sensitive protection than traditional overcurrent protection, as negative sequence currents are only present during fault or transient conditions, therefore the relay does not have to allow for full load current.

4.6. Specific applications

4.6.1 Overview

A multitude of specialized relays has been developed over the years for specific applications. Generator protection, for instance, demanded additional functions like field failure protection, loss of excitation, stator earth fault, rotor earth fault, etc. It is not the purpose of this text to discuss all these specific applications in detail, but some reference to the more important requirements are made in the next section, mainly to highlight the versatility and protection power of today's multi-function relays.

Electromechanical relays could only perform one function per relay, and their characteristics were inflexible, for example a relay could only have one time–current curve characteristic (normal inverse, very inverse, etc). Generally, it was only their settings that could be varied. Anything more meant replacing the relay, often with extensive panel wiring modifications.

This also meant that, for example, a generator protection panel would have a multitude of relays, one relay for each protection function.

This changed dramatically with the advent of digital relays, in two ways:

First, the relays became flexible. Different curve characteristics, for instance, were programmed into the relay during manufacture and were user selectable, usually through dipswitches or pushbuttons on the first digital relays. Then communication capabilities were developed, which meant that more of the relay configuration could be downloaded to the relay after installation, increasing flexibility tremendously. Relay processing power developed along with the computer industry, and today's digital relays offer not only selectable curve characteristics, but up to eight different complete protection setting selections, selectable automatically by an external input or an event occurrence. This is very useful, especially in applications like ring supplies, automatic changeovers, etc.

Furthermore, some relays offer the opportunity to the user to construct his own time–current curve, superimposing normal-, very- and extremely inverse characteristics to form one curve. The most advanced relays offer 15 different curves, including two that can be user-configured, superimposing any of the 13 other curves.

Adding to flexibility came versatility with digital relays. This means that more than one function can be incorporated in the same relay. The first digital relays incorporated the well-known and widely used principle of two overcurrent relays and one earth-fault relay into one relay. This concept is still widely used in the basic level digital relays offered by most manufacturers, and feeder protection relays. As microprocessor power developed, so did multi-functionality of relays. The ultimate example today is a generator protection relay, which can typically incorporate up to 32 complex protection functions in one relay.

Thereafter communication capabilities of relays developed, and eventually control functions were introduced. This created a device that is more than just a relay, more than just a communication device, more than a control device, and the term IED (intelligent electronic device) was introduced to describe these modern devices. IEDs, their applications, and their critical role in power system automation systems are discussed further in Chapter 5.

Initially, relays were classified according to their function, for example an overcurrent relay or an earth fault relay. This is still the case for the more basic types of relays, but today's intelligent relays and IEDs are rather classified according to their application, as most are multi-function, for example a generator protection relay, transformer protection relay, feeder protection relay, etc.

Most multi-function relays today include extensive communication capabilities for incorporation in a power system automation system.

4.6.2 Generator protection

Most leading manufacturers of relay equipment today have a multi-function generator relay on the market, offering a host of functions. The relay has to be configured to meet the customer's requirements and according to the equipment specifications. All of the functions available need not be required and hence will not be configured. Configuration of complex multi-function relays like a generator protection relay is quite involved and is usually done by the relay supplier, preferably on a factory test-set before installation.

Configuration costs can be quite high, depending on the amount of functions required. Therefore, the manufacturer usually quotes a certain fixed price for the hardware, and additional configuration costs according to the application.

Important factors for reliable generator operation are the following:

- Maintaining a magnetic field in the stator: Without this magnetic field, which is called excitation, an emf cannot be setup. The generated emf causes a braking force on the rotor, which ensures that a stable speed is maintained. When excitation is lost, this braking force is not present, and if high-pressure steam is still applied to the rotor, the rotor may accelerate to unacceptably high speeds, causing mechanical damage to the generator. Electrical instability of the generator will also result, with very detrimental effects on the power network.
- Therefore, a 'loss of excitation' function is employed, which will ensure that the source of mechanical power to the generator (e.g., high-pressure steam) is shut down when excitation is lost.
- The generator has to be protected against reverse power being applied to it, i.e. electric power from another source. This may occur when excitation has been lost to the generator, and a potential is applied to the generator while the rotor is still turning due to momentum. This will cause the generator to behave like a motor, and the applied potential may attempt to turn the rotor in the other direction, causing too high mechanical stress and rotor damage. This is called reverse power protection.
- In addition, the generator is protected against electrical faults which may occur in the generator, like stator earth fault and rotor earth fault, and faults that may occur in the distribution network, like short-circuits and earth faults. The generator will be protected against overloading by overcurrent protection, and the distribution network will be protected against the generator malfunctioning, for example deviations like under- and overfrequency, under- and over-voltage, etc.

4.6.3 Transformer protection

The functions of transformer protection relays are less complex than those of generator protection, and the configuration of the relays is more standard, depending mostly on the equipment specifications. Generally, differential protection will be applied, with overcurrent and earth fault functions as a backup. Larger transformers may also have frequency protection.

Supplementary protection devices will usually be interfaced to the transformer, e.g. Buchholz protection, oil temperature, etc.

4.6.4 Feeder protection

Feeder protection can be applied in two forms: Feeder differential protection, as discussed in section 4.5.3, or the traditional overcurrent and earth fault functions, which are undoubtedly the most commonly used form of protection.

An overcurrent and earth fault relay (in recent times only one relay) will generally be installed on each feeder cubicle in a distribution substation, often supporting or as a back-up for other specific types of protection, like differential protection or transformer protection.

The feeder protection relay (differential protection is for now excluded from the definition) functions independently and in a very straightforward manner. It is this characteristic, together with the fact that it is so commonly used, that made this relay the ideal candidate to be developed into a versatile, flexible intelligent relay, with powerful control functions and advanced communications capabilities. For example, the supplier ABB has named their feeder protection relay a feeder terminal, as it includes much more than only feeder protection functions.

This is discussed further in Chapter 11.

4.6.5 Busbar protection

The sole function of busbar protection is to protect the busbars of a switchgear panel against internal faults, that is, faults within the clearly defined busbar zone.

Busbar faults are by nature quite severe, with high fault currents flowing. The damage to the switchgear panel, which is normally a very expensive piece of equipment as well as a crucial part of the distribution network, can be extensive if the fault is not cleared quickly. Therefore, busbar protection needs to operate very fast and reliably.

Figures 4.12(a) and (b) illustrate a typical distribution switchgear panel with zones of protection. Only one zone of protection may be defined, as in Figure 4.12(a). This would be the lowest cost option but with no fault discrimination. Alternatively, different zones may be defined, as in Figure 4.12(b). This is a more costly option, but fault discrimination is obtained.

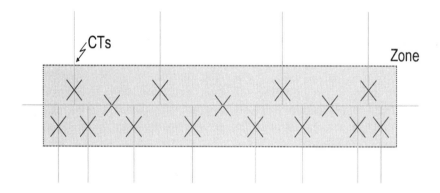

Figure 4.12(a)
Single zone busbar protection

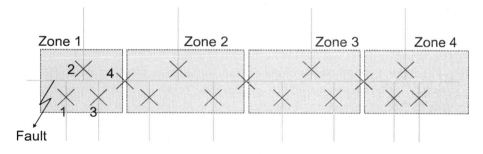

Figure 4.12(b)
Multiple zone busbar protection

Normally, with a busbar fault, all circuit breakers of that panel are tripped, including the feeders. This is to prevent any large synchronous motor acting as a generator and feeding into the fault while the rotor still has mechanical momentum. Achieving fault discrimination as in Figure 4.12(b), only circuit breakers nos 1–4 need to be tripped, removing the faulted part from the rest of the panel.

Three methods of busbar protection are generally used:

Busbar zone protection

Busbar zone protection is basically an extended version of current differential protection. The current of each and every bay of the switchgear panel is measured, totaled and a fault detected if the vector sum of all the currents does not add up to zero, after compensating for internal losses, CT and relay inaccuracies.

Busbar zone protection is usually classified as either high-impedance or low-impedance zone protection, the main difference is the type of equipment being used, resulting in significant difference in speed of operation and cost.

Low-impedance zone protection is the more expensive and operating speeds in the region of 40–60 ms are generally obtained.

High-impedance zone protection is less expensive, but slower operating speeds in the region of 100–150 ms are generally obtained. However, this is still a lot faster than overcurrent or earth fault protection.

Arc protection

This is a recent technology, and offers quite an economic method of busbar protection, with high speed of operation (in the region of 80 ms).

Arc protection operates on the principle that an arc, giving off intense light, will be present in the busbar chamber right from the start of a busbar fault. This light flash is detected and the arc protection relay immediately gives the tripping command.

Arc protection schemes usually offer the options of either operating solely on the detection of light, or to combine the light detection with overcurrent relays on the incomers. The arc protection will then operate only when a flash of light is detected, combined with the sensing of overcurrent, increasing stability of the scheme, while not compromising on reliability.

The overcurrent setting will be relatively low, just above full load current, so as not to introduce a significant time delay. The arc protection will not operate on overcurrent alone, therefore no allowance has to be made for normal transients like motor starting; and grading with other relays is not required.

Two methods are currently used for detection of the arc. One method consists of looping a bare fiber optic through the busbar chamber (one fiber optic for each zone). This is the lower cost option, and also the less 'intelligent' one, as the fiber optic detects the arc, but the relay does not know where it came from. However, this method provides fast, reliable protection for the busbar chamber(s).

The second, more costly but also more 'intelligent', option involves the installation of discrete light sensors to detect an arc. The relay will then know exactly from which sensor the arc detection originated, and a high level of discrimination may then be obtained. This method is especially suited to provide fast protection in the cable chamber of switchgear panels, which normally falls outside the protective zone of busbar protection. This is a relatively high-risk area for faults, due to the possibility of cable insulation being damaged during termination. Generally this area is only protected by overcurrent and earth fault protection, except if feeder differential protection is used.

Busbar blocking

This is a relatively low cost, very effective type of busbar protection, although slower operating speeds are obtained compared to busbar zone protection and arc protection. The principle of operation is quite ingenuous, although very straightforward, and has very flexible possibilities when using intelligent relays.

This method is illustrated in Figure 4.13. The principle of operation is as follows:

Overcurrent relays are used on all incomers and feeders. These relays still perform their normal functions of providing overcurrent protection for their zone, as well as participating in the busbar-blocking scheme. The relays will be set to identify a *possible abnormal condition* slightly above normal full load current, i.e. a relatively sensitive overcurrent detection.

Referring to Figure 4.13, three possible scenarios may occur:

- Scenario 1: A 'normal' system transient, for example a large motor starting. Relay 1 and Relay 2 will both detect the overcurrent. Relay 1 will send a 'blocking signal' to Relay 2, ensuring that Relay 2 does not trip. The time setting of Relay 1 needs to be such that the transient will pass before the relay sends a tripping signal to the circuit-breaker. (Depending on the size and load characteristic of the switchgear panel, Relay 2 may not even detect the motor starting, as the full load current setting of Relay 2 needs to allow for the load of the whole panel.)
- Scenario 2: A phase-to-phase fault at point A. Relay 1 and 2 will both detect the overcurrent, Relay 1 will send a blocking signal to Relay 2, and according to the time–current settings, Relay 1 will send a tripping signal to the circuit-breaker, clearing the fault.
- Scenario 3: A phase-to-phase fault at point B. Relay 1 will not detect the overcurrent, nor any other feeder protection relay, no blocking signal will be sent to Relay 2, and Relay 2 will trip the instant the overcurrent is sensed.

This time setting can be very short, only allowing time for a possible blocking signal to arrive, as Relay 2 need not compensate for motor starting currents, or grade with Relay 1.

To summarize, any feeder protection relay in a busbar blocking scheme that detects an overcurrent will send a blocking signal to the incomer relay(s). The incomer relay(s) will trip (almost) immediately when an overcurrent is detected, *but only in the absence of a blocking signal*. Thus proper discrimination is obtained.

A few points worth noting:

- The feeder protection relays must be able to send a signal to the incomer relay(s), and the incomer relay(s) must be able to receive a trip inhibit signal.
- This scheme is slower than busbar zone protection or arc protection, because the incomer relay(s) first have to detect the overcurrent. Relay tripping times in the order of 200 ms can normally be obtained.

The same principle can be applied in other protection schemes as well, eliminating the need for grading between relays.

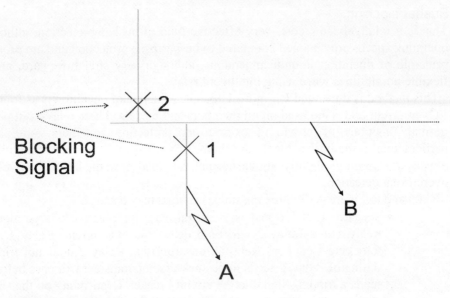

Figure 14.13
Illustration of busbar blocking

4.7 Comparison of electromechanical relays and digital relays

No	Feature	Electromechanical	Digital
1	Reliability	Moderate	High
2	Stability	High	High
3	Sensitivity/Accuracy	Low	High
4	Speed of operation	Moderate	High
5	Discrimination capability	Moderate	High
6	Multi-function	No	Yes
7	Versatile (can be used for different applications)	No	Yes
8	Flexible (multiple curves, selectable setting groups)	No	Yes
9	Maintenance intensive	High	Low
10	Self-diagnostics	No	Yes
11	Trip circuit supervision	No	Yes
12	Condition monitoring	No	Yes
13	Data communications	No	Yes
14	Control functions	No	Yes
15	Metering	No	Yes
16	Disturbance recordings	No	Yes
17	Remote operation	No	Yes
18	CT Burden	High	Very low
19	Cost	Low	See note 1

Note 1:

The cost will depend on the application. If a multi-function relay, like an intelligent feeder terminal, is used for only overcurrent protection, the cost for that function will be high. However, if a digital relay is fully used, for example, a generator protection relay, the cost per function will be low.

Note 2:

Not all digital relays will comply with the features as in the table, depending on the type of relay. However, the table is an overview of what is possible with digital relays in general, compared to what could be achieved with electromechanical relays.

The tremendous capabilities of microprocessor-based digital relays are comprehensively described in the section on IEDs in the next chapter.

5

Remote substation access and local intelligence

5.1 Introduction

Remote substation access as it is utilized today, that is, remote control and monitoring of substations from a central control room, actually developed from different industries, with different goals, perceptions and philosophies. That is the reason different architectures exist today in power system automation. Undoubtedly the future will see increasingly uniform systems, operating on the same principles with the same philosophies, emerging, and being applied. However, it is important to understand the background of these developments, in order to recognize and appreciate the different emphasis and strengths of the various power system automation systems available today.

5.2 The remote terminal unit

The remote terminal unit (RTU) was developed with the aim of acting as an interface and communication unit between field instruments and a SCADA master. The first RTUs did little more than converting the information from the instruments into the SCADA language, send the information through, and vice versa. RTUs were primarily developed, and continually improved, with strong communication capabilities in mind, as this was their purpose.

The electrical utility industries first use RTUs in large substations to collect a vast amount of digital and analog data, such as alarms, events and measurements, and to send them through to the SCADA station, where they could be viewed by the operator, acted upon and/or stored for historical purposes. Remote, or centralized, control was achieved by huge mimic panels.

RTUs developed to become more intelligent and local control functions were used increasingly on these devices, restricting the chances of human error and the dependability on experienced operators. Local control functions were incorporated to eliminate the need to rely on the master station, moving closer to the PLC philosophy. The terms ERTU (enhanced RTU) and smart RTU came into existence to describe these devices.

This has advanced to the stage where virtually the complete substation's control functions would be incorporated into one large, powerful RTU with an interface to an advanced SCADA station. This has the disadvantage that the reliable operation of the substation is wholly dependent on one device, but substation RTUs have been developed with reliability and integrity as core issues, and the equipment from leading manufacturers seldom experience a catastrophic failure. Nevertheless, failures can and do occur, and critical substations are often provided with redundant RTUs, adding greatly to the cost of the substation.

Substation RTUs are by nature very costly equipment; therefore their use is economically justified only in transmission substations, where potential loss of revenue far outweighs their expense. The high cost of substation RTUs has restricted their use in distribution substations.

Currently, RTUs are used in an estimated 80% of transmission substations and 40% of distribution substations worldwide.

The huge amount of wiring required to a central RTU, has initiated the developments of small RTUs, designed to be installed close to the switchgear and communicate to the master RTU. This required the RTUs to communicate to each other, sparking the concept of communications within a substation environment.

A small step from here enabled small RTUs to communicate directly to the SCADA station, and this development has made RTUs an affordable and justified option for distribution substations. Small RTUs have subsequently been developed with the main purpose of controlling one or two bays of a distribution substation, and these devices have acquired the name of 'bay controllers'.

5.3 Programmable logic controller

The programmable logic controller (PLC) has been developed specifically for process automation. However, the versatility, ease of application and relatively low cost of PLCs compared to RTUs has facilitated their use in many industrial substations, especially in North America.

The major disadvantage of using PLCs in a substation environment is their inherent weakness of communicating to relays. Some manufacturers have gone this route, however, developing PLC-based power system automation systems, enhancing the relay-PLC communications of their systems.

The logic functionality of PLCs has been duplicated into the advanced relays of certain manufacturers, creating relays which have the added capability of performing logic functions, hence local control functionality, giving rise to the term 'intelligent relays'.

5.4 Protection relays

As discussed earlier, protection relays became more advanced, versatile and flexible with the introduction of microprocessor-based relays. The initial communication capabilities of relays were intended mainly to facilitate commissioning. Protection engineers realized the advantages of remotely programming relays, the need developed for data retrieval, and so the communication aspects of relays became steadily more advanced.

PLC functionality became incorporated into relays, and with the development of small RTUs, it was soon realized that relays can be much more than only protection devices.

Why not equip protection relays with advanced control functions? Why not add protection functions to a bay controller? Both of these approaches have been followed, with different manufacturers (and sometimes different divisions within the same manufacturing group) adopting different approaches to the question of protection, control, and communications. This resulted in an extensive range of devices on the market, some stronger on protection, some stronger on control, and the term protection relay became too restrictive to describe these devices. This resulted in the term 'intelligent electronic device' (IED).

5.5 The intelligent electronic device (IED)

5.5.1 Definition

The term 'intelligent electronic device' (IED) is not a clear-cut definition, as for example the term 'protection relay' is. Broadly speaking, any electronic device that possesses some kind of local intelligence can be called an IED. However, concerning specifically the protection and power system automation industry, the term really came into existence to describe a device that has versatile electrical protection functions, advanced local control intelligence, monitoring abilities and the capability of extensive communications directly to a SCADA system. This is the definition of an IED that will be applied throughout this book.

A multitude of relays from different manufacturers can perform the functions of protection, control, and monitoring (including measurement), but need the assistance of an RTU or communications processor, to which they are hardwired, to communicate with the SCADA supervisor. These devices may be called intelligent relays but are not included in the definition of an IED.

Similarly, some relays can communicate directly to a SCADA, but lack the control functionality. These relays are often used in conjunction with bay controllers, which provide the required control functions, to form a power system automation system. Again, these relays cannot be classified as IEDs.

The ability of an IED to perform all the functions of protection, control, monitoring, and upper level communications independently and without the aid of other devices like an RTU or communications processor (not including interface modules) is the identifying feature of an IED.

Note: The above definition refers to ability, and not to a specific application, where an IED may well communicate to an RTU, for example.

5.5.2 Functions of an IED

The functions of a typical IED can be classified into five main areas, namely protection, control, monitoring, metering and communications. Some IEDs may be more advanced than others, and some may emphasize certain functional aspects over others, but these main functionalities should be incorporated to a greater or lesser degree.

5.5.3 Protection

The protection functions of the IED evolved from the basic overcurrent and earth fault protection functions of the feeder protection relay (hence certain manufacturers named their IEDs 'feeder terminals'). This is because a feeder protection relay is used on almost all cubicles of a typical distribution switchboard, and the fact that more demanding protection functions are not required enable the relay's microprocessor to be used for control functions. The IED is also meant to be as versatile as possible, and is not intended to be a specialized protection relay, for example generator protection. This also makes the IED affordable.

The following is a guideline of protection related functions that may be expected from the most advanced IEDs (the list is not all-inclusive, and some IEDs may not have all the functions). The protection functions are typically provided in discrete function blocks, which are activated and programmed independently.

- Non-directional three-phase overcurrent (low-set, high-set and instantaneous function blocks, with low-set selectable as long time-, normal-, very-, or extremely inverse, or definite time)
- Non-directional earth fault protection (low-set, high-set and instantaneous function blocks)
- Directional three-phase overcurrent (low-set, high-set and instantaneous function blocks, with low-set selectable as long time-, normal-, very-, or extremely inverse, or definite time)
- Directional earth fault protection (low-set, high-set and instantaneous function blocks)
- Phase discontinuity protection
- Three-phase over-voltage protection
- Residual over-voltage protection
- Three-phase under-voltage protection
- Three-phase transformer inrush/motor startup current detector

5.5.4 Control

Control functions include local and remote control, and are fully programmable.

- Local and remote control of up to twelve switching objects (open/close commands for circuit-breakers, isolators, etc)
- Control sequencing
- Bay level interlocking 1 (one?) of the controlled devices
  – Status information 2
  – Information of alarm channels 2
- HMI panel on device

Notes:
1. Secure station level (interbay) interlocking demands peer-to-peer communications of <10 ms. This is not supported by all manufacturers' systems, as will be seen in Chapter 12.
2. Status and alarm information is part of the control function blocks as they have a direct bearing on secure control functions.

5.5.5 Monitoring

Monitoring includes the following functions:

- Circuit-breaker condition monitoring, including operation time counter, electric wear, breaker travel time, scheduled maintenance
- Trip circuit supervision
- Internal self-supervision
- Gas density monitoring (for SF6 switchgear)

- Event recording
- Other monitoring functions, like auxiliary power, relay temperature, etc.

5.5.6 Metering

Metering functions include:
- Three-phase currents
- Neutral current
- Three-phase voltages
- Residual voltage
- Frequency
- Active Power
- Reactive Power
- Power Factor
- Energy
- Harmonics
- Transient disturbance recorder (up to 16 analog channels)
- Up to 12 analog channels

5.5.7 Communications

Communication capability of an IED is one of the most important aspects of power system automation and is also the one aspect that clearly separates the different manufacturers' devices from one another regarding their level of functionality.

By definition, IEDs are able to communicate directly to a SCADA system, i.e. upper level communications. Different manufacturers use different communication protocols, which will be discussed in detail in the next section.

An IED will, in addition to upper level communications, also have a serial port or optical interface to communicate directly to substation PC or laptop for configuration and data downloading purposes, should the SCADA link not be available or desirable in that instance.

6

Data communications

6.1 Basic requirements of communication

The basic requirements of any communication are the following:

- Physical link: There can be no communication without a physical link between the sender and receiver, e.g., a radio link, telephone link, etc.
- Agreed medium: The medium of communication must be the same, i.e., voice communication or data communication, etc. For example, two persons can be linked through a telephone wire, but the sender is sending a fax while the receiver is expecting a voice message. No communication is taking place.
- Same language: No communication is taking place if, for example, the sender is speaking French while the receiver understands only English.
- Common context: For example, the (English) word 'lie' can have vastly different meanings, depending on the context in which it is spoken.
- Receiver identified: The receiver(s) of the message must be clearly identified if there is more than one possible receiver on the link (except in the case of 'open' one way communication, e.g., a television broadcast).

Without the above basic requirements, there can be no meaningful communication. Data communication is much more complex than this, but still shares the same basic principles.

Data communications in the context of power system automation refer to digital data, i.e., information represented by a sequence of zeros and ones. Many communications systems handle analog data, for example telephone, radio and television, although even in these systems digital methods are gaining ground fast (e.g., digital satellite television).

Digital transmission is superior to analog signals due to enormous flexibility, more information capacity, and virtual noise immunity.

Digital data is often transferred over links that were designed for analog communication, e.g., computer communications over a telephone line. A modem is then used to modulate the digital data onto an analog waveform, and at the other end a modem then demodulates the waveform to reproduce the original digital data. (The word 'modem' was derived from 'modulator'/'demodulator'.)

The reverse also often occurs. An analog waveform (e.g., sound) is 'coded' into a digital signal, sent over a digital link (e.g., digital satellite) and 'decoded' into the original waveform at the receiver.

Once the physical link and communication medium have been established, e.g., data communication over a fiber optic link, the language, context, and identifying aspects need to be defined. The way this is done in data communications is called a 'protocol'.

Communication protocols are discussed in detail in Chapter 7.

Note: The term 'communication' will exclusively refer to digital data communication for the rest of the text, unless otherwise stated.

6.2 Communication topologies

Three communication topologies, each with different variations, are mainly being used for network applications, namely star-, ring-, and bus configurations. The topology may refer to the physical topology (i.e., the wiring layout of a network) or the logical topology (i.e., how information is functionally transmitted through the network). The two topologies may not necessarily correspond in any given network, and often combinations of topologies are used.

The following discussion focuses on logical, rather than physical topologies.

6.2.1 Star topology

This configuration is illustrated in Figure 6.1. Multiple stations or nodes are connected to a central component, known as the master (logical) or hub (physical).

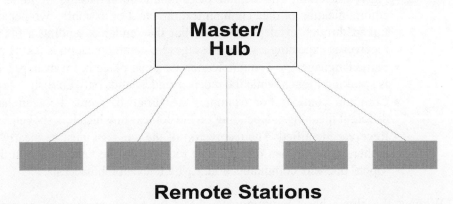

Figure 6.1
Star topology

Advantages

- Straightforward and easy to maintain
- Easy troubleshooting and fault isolation
- Easy addition or removal of nodes
- Failure of a single node does not influence the network
- Easy monitoring of data traffic for management purposes

Disadvantages

- The entire network is dependent on the hub (master). If it fails, the whole network is down. Often a redundant central station is provided for this eventuality.
- Direct communications between nodes are not supported, everything must go through the hub.

6.2.2 Ring topology

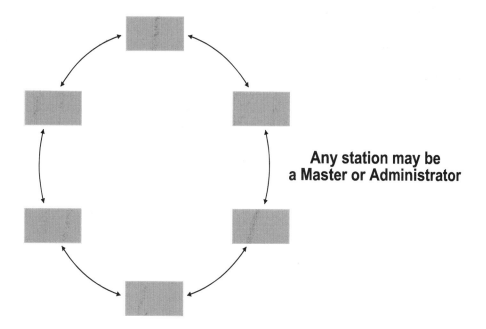

Figure 6.2
Ring topology

The ring type configuration is illustrated in Figure 6.2. Message packets are transmitted sequentially from node to node in a predefined sequence. A closed loop is formed with information travelling one-way.

Advantages

- The network is not necessarily dependent on a master, depending on the technique used, although an administrator may be required with certain techniques.
- Each node can function as a signal amplifier.
- Automatic message acknowledgement is inherent to the topology (if no station accepts the message, it automatically returns to the sender).
- Direct node-to-node communication is supported.

Disadvantages

- If any node goes down, the entire ring goes down. Later developments served to alleviate this drawback by introducing a double-ring concept, where data can circulate in both directions around the ring. Therefore, the ring can still function with one node out of service. The ring will break down when a second node fails.
- Troubleshooting and fault isolation are difficult.
- Adding or removing nodes disrupts the network.
- Configuration and programming of the system become more complex.

6.2.3 Bus topology

The bus topology is illustrated in Figure 6.3. Single bus networks, double (redundant) bus networks, and complex combinations are used in practice. The 'bus' refers to the main communication channel to which each node is connected. Every node listens to the bus and checks the destination address of transmitted messages, accepting the message if it is intended for the node. Messages travel in both directions on the bus and do not need to go through individual nodes. If no node accepts a message, an electrical terminator at the end of the bus absorbs the energy to keep the message from reflecting back along the bus, possibly interfering with other messages sent subsequently on the bus.

This is the most flexible of the different topologies. Any communication technique can be used on a communication bus. The bus need not be dependent on a master or a bus administrator, depending on the technique used.

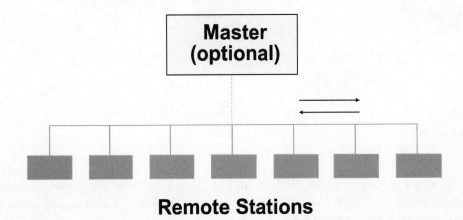

Figure 6.3
Bus topology

Advantages

- The bus is not dependent on a single central station
- Any one node may fail without affecting the bus
- High flexibility of configurations
- Nodes can be added or removed from the bus easily
- Direct node-to-node communication is supported

Disadvantages
- Security may be compromised, as any node can see the message.
- Troubleshooting and fault isolation can be difficult, since the fault can be anywhere on the bus.
- There is no inherent automatic acknowledgement of messages, since messages are absorbed at the end of the bus, and do not return to the sender.
- Heavy data traffic on the bus may cause problems, as nodes may find it difficult to access the bus to send a message.

6.3 Communication techniques

Communication techniques are divided into two types: master-slave communication and peer-to-peer communication.

6.3.1 Master–slave communication

This is the simplest, most common, but also least flexible communication technique. Centralized control over the communication network is executed by the 'master', normally a SCADA system. The rest of the devices on the communication network are the 'slaves', and may only communicate on the network when requested or permitted by the master, and only with the master.

Master–slave communications usually result in inefficient use of communication bandwidth, and relatively slow speed of communication; although many systems allow priorities to be assigned to send/gather data more quickly to/from certain specified sites.

Deterministic response times are obtained with this technique. It can be used on any topology.

6.3.2 Peer-to-peer communication

The peer-to-peer communication technique allows all devices on the network to initiate communications with any other device on the network.

Although there is not a master and slave configuration, some techniques require a bus administrator to assume a certain amount of communication control over the network. One of the remote stations will then perform this role. Other techniques do not require a centralized communication controller and each device on the network will perform its own communication control. This demands far more complex protocols, but more flexibility, efficient use of bandwidth and hence far higher speeds of communication are obtained.

Although it is not functioning as a master, a SCADA system will normally receive the majority of the network data, and remote control commands will usually be initiated by the SCADA system. Although supervisory control and data acquisition are lost if the SCADA system should go down, network communications can still be retained (which is not the case in master–slave techniques).

By nature, peer-to-peer communications cannot be used on a star topology.

6.4 Media access control principles

'Rules' are needed for data communications, otherwise there will be chaos, and no communication can exist. These rules govern the way stations will get access to the network, thus the term 'media access control'.

6.4.1 Conventional polling

Conventional polling is the most common and simplest form of data communication. A master station requests data from each slave station in a predetermined sequence. If a slave does not respond in a defined time, the master retries (the number of retries depending on the programming), and marks the slave as non-responding if still no response is received.

Polling is illustrated in Figure 6.4.

Priorities may be assigned to slaves, meaning these stations will be polled at a higher rate than the normal stations. (Naturally, making all stations priority stations will defeat the purpose.) A priority message sent from the master station can override the standard polling sequence.

Polling can be used on virtually all physical media and any topology. The polling rate is dependent on the master and deterministic response times are obtained.

Polling results in inefficient use of bandwidth and relatively slow response times are characteristic of polling techniques.

No peer-to-peer communications are possible with polling.
(Note: Also commonly called 'cyclic polling'.)

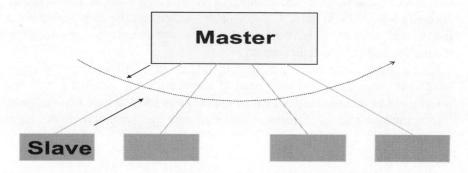

Figure 6.4
Polling technique

Advantages of polling

- Ease of programming
- Reliability due to the simplicity of the philosophy
- Failure of communication, either at the slave or due to a link failure, is detected relatively quickly.
- No data collisions can occur on the network, therefore the data throughput is predictable and constant.
- An efficient and predictable system for heavily loaded systems with each slave having constant data transfer requirements.

Disadvantages of polling

- Systems that are lightly loaded with minimum data changes from slaves are inefficient and unnecessarily slow.
- Slaves cannot communicate directly with each other, but have to do so through the master, adding complexity to the master programming, and resulting in slow communications between slaves due to the polling sequence.

- Variations in the data transfer requirements of slaves cannot be handled as these are defined during programming.
- Interrupt type requests from slaves requesting urgent action cannot be handled, as the master may be processing some other station.

6.4.2 Polling by exception

While maintaining most of the characteristics of conventional polling, this technique consists of the master requesting only event changes from the slave stations. If no changes have occurred during the polling cycle, no data will be returned to the master. This is a much more efficient data gathering technique for stations that have large amounts of data that vary infrequently.

Note that the protocol used in communicating between master and slave stations must support polling by exception, and both master and slave must understand the concept of events versus static data. Events are time-tagged at the time of their occurrence, which is quite an improvement over conventional polling where the time of an event occurrence may be inaccurate by as much as the polling cycle time.

6.4.3 Time-division-multiplex media access

With time-division-multiplexing (TDM), each station has its own timeslot, during which the station may send and receive data. This can be illustrated as in Figure 6.5. The time slot is fixed, independent of the network load. Deterministic response times are achieved, as there can be no collisions on the network.

A bus administrator is required to ensure that all stations comply with the network 'rules', also called centralized transport media arbitration.

TDM can be used on a ring or bus topology.

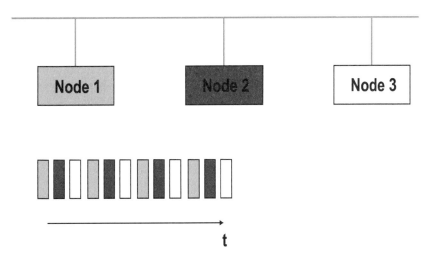

Figure 6.5
Illustration of TDM

Advantages of TDM

- No data collisions can occur on the network, therefore the data throughput is predictable and constant.
- It results in an efficient and predictable system for heavily loaded systems with each station having constant data transfer requirements.
- Direct peer-to-peer communications are possible, although this is still relatively slow due to each station having to wait for its timeslot.

Disadvantages of TDM

- Systems that are lightly loaded with minimum data changes are inefficient and unnecessarily slow.
- Variations in data transfer requirements cannot be handled as these are defined during programming.
- Interrupt type requests from stations requesting urgent action cannot be handled.
- A communication failure to a specific device may not be detected immediately, except in protocols that use a technique called background polling or integrity polling. The bus administrator then polls each station (at a much slower rate) to verify that the central station's data is up-to-date and to check the health of stations.
- The network is still dependent on a central communications controller.
- The configuration of the network is quite complex.

6.4.4 Token passing media access

The principle of this technique is illustrated in Figure 6.6. A token circulates around the ring. Any station that has something to report may take the token as it passes by, and that station then has exclusive access to the communication network. Once the station has completed its transmission, the token is released and circulates around the network again. This technique generally, but not necessarily, operates on a report-by-exception principle.

Semi-deterministic response times are obtained, as there can be no data collisions, but it cannot be pre-determined when a station is going to use the token or simply pass it on.

A centralized bus administrator is required to control the network communications.

Protocols that use this principle are either token ring or token bus protocols. Token ring protocols use a physical ring and messages pass through each station as it circulates around the ring. With token bus protocols, the token still circulates in a logical ring, but the physical network is a bus topology, with the result that failure of one station will not necessarily affect the network.

Advantages of token passing

- No data collisions can occur on the network.
- Direct peer-to-peer communications are possible, faster than any of the previous techniques, with each station only having to wait for the token to pass by or be released by another station.
- It results in a more efficient system for lightly loaded systems with each station having variable data transfer requirements.
- Variations in data transfer requirements of the stations can be handled by the system.

Disadvantages of token passing

- A communication failure to a specific device will only be detected when a specific response was requested from a station, and none was submitted. (This can be alleviated by background polling.)
- The network is still dependent on a central communications controller.
- The configuration of the network is quite complex.
- Only semi-deterministic response times are obtained.
- Unnecessary waiting times are still inherent to the technique (stations waiting for the token to pass).

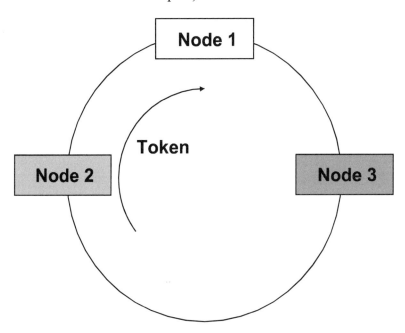

Figure 6.6
Illustration of token passing

6.4.5 CSMA/CD (carrier sense multiple access with collision detection)

With this technique, each station listens to the communication bus (multiple access), and sends when it has detected that the bus is idle ('carrier sense'). If a collision occurs with another station, the transmission will be stopped, and after a random period the station will retry (collision detection).

All stations must be capable of supporting collision detection, collision avoidance, and recovery schemes. Methods for these vary considerably and thus have become a major issue in compatibility between different systems.

This technique is very efficient in maximizing the use of available bandwidth, and offers very flexible configuration options. It was designed to be used in a bus topology.

Response times are undeterministic, as network data load cannot be predetermined.

Ethernet communications are based on this technique.

CSMA/CD is illustrated in Figure 6.7.

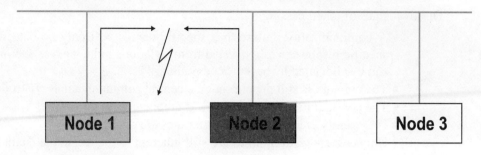

Figure 6.7
Illustration of CSMA/CD

Advantages of CSMA/CD

- Direct peer-to-peer communications are possible, faster than any of the previous techniques. This is the most suitable technique for peer-to-peer communications.
- It results in a very efficient system for lightly loaded systems as well as still being efficient for heavily loaded systems.
- Variations in data transfer requirements of the stations can be handled by the system.
- Urgent requests for network access from a station can be handled instantly, except if another station already has access. Some protocols have priority access techniques.
- No centralized bus controller is required.

Disadvantages of CSMA/CD

- A communication failure to a specific device will only be detected when a specific response was requested from a station, and none was submitted (except if a SCADA system or centralized controller do background polling.)
- The configuration of the network is very complex.
- Non-deterministic response times are obtained.
- Data collisions are an inherent drawback of the technique, requiring very involved data collision detection, avoidance, and recovery schemes.

6.5 The OSI model

6.5.1 Basic principles of transmitting data

Digital data is transmitted by using a series of zeros and ones, which are called bits. One character (alphabetical, numerical or other character like a question mark) is defined by a series of bits (usually eight, depending on the protocol), called a byte. A series of bytes are sent in a frame, which contains the message. This is the raw data that has to be transmitted. The principle is illustrated in Figure 6.8. In the most basic master–slave communications, the message can be sent without adding a lot of information, possibly only adding time-stamping, error detection, and start/stop bits.

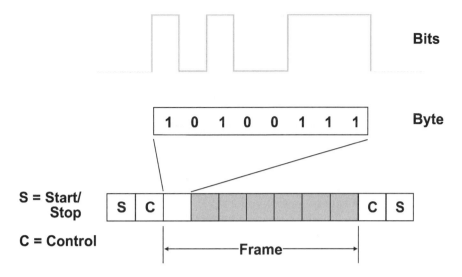

Figure 6.8
Data frame construction

However, in complex bus topologies with peer-to-peer communications, quite a lot of information needs to be added to the raw data, for example the address, length of message, time stamping, more advanced error detection and correction information, communication control information, protocol information, context information, etc.

The need for all this information to be added to the raw message can, and did, lead to a lot of confusion and incompatibility between different devices on the network.

The open system interconnection (OSI) model was developed by the International Standards Organisation (ISO) to provide a universally applicable structure for communication applications.

6.5.2 The OSI model

The OSI model consists of a seven-layer stack onto which the message (raw data) is modulated. The way in which the seven layers is applied (not all seven layers necessarily need to be used) actually defines the communication protocol. The OSI stack is illustrated in Figure 6.9.

Application Process

7	Application
6	Presentation
5	Session
4	Transport
3	Network
2	Data Link
1	Physical

Figure 6.9.
Construction of OSI seven-layer stack

The concepts of the OSI layers are as follows. Each layer:
- Performs unique and specific tasks
- Provides service to the adjacent layer above
- Uses the service of the adjacent layer below
- Has knowledge of only its immediate adjacent layers

The functions of each layer are subsequently discussed, referring to Figure 6.9.

Physical layer
This layer provides the mechanical characteristics of the physical interface (the medium, e.g., copper, fiber optic, and the connectors), as well as the electrical characteristics (level and frequency of electrical/optical pulses).

Furthermore, the functional and procedural characteristics to activate, maintain, and deactivate the connection are defined.

Data link layer
The data link layer provides the synchronization (start/stop, length) and error control (detection and correction) for the information transmitted over the physical link. The information provided by the physical layer is used in defining this layer.

Network layer
This layer provides the means to establish, maintain, and terminate the connections between stations, including routing and address functions.

Transport layer
The transport layer provides end-to-end control and information interchange with the level of reliability that is required for the application (e.g., confirming connections, sequencing, etc). This layer functions to liaise between the end-user(s) and the network.

Session layer
This layer supports the dialog requirements on the bus, determining who is allowed to talk, handling start/stop talking commands and communication relationships. In this context, application processes (AP) are defined, rather than physical stations.

Presentation layer
The presentation layer provides the interpretation and meaning of the information exchanged, in other words the coding or 'language' of the information.

Application layer
This layer directly serves the end-user, the AP, by providing the information to support the AP and manage the communication. This layer thus specifically defines the information and its context to the AP.

The specific application can be said to be directly on top of the application layer.

6.5.3 Communication using the stack

Figure 6.10 illustrates the way each layer of the OSI stack will add information to the raw data at the sender's end, and strip away the added information at the receiver's end to again arrive at the raw data, which is the actual required message.

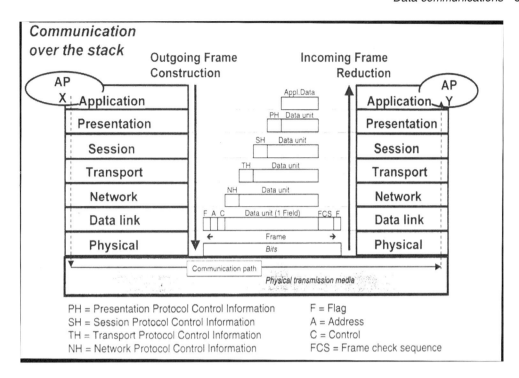

Figure 6.10
Movement of data through the stack

A basic example will illustrate the process. (Note that this example is not the absolute rule, as every protocol will differ to a certain extent in the specific application of the layers. The example does illustrate the process, however.)

Example

A voltage value needs to be sent over a network. The value will be encoded into digital format by an analog-to-digital converter, which is not part of the communications network. The AP in this instance is voltage measurement.

As Figure 6.10 shows, data will move down the OSI stack, each layer adding information to it. The application layer will add meaning to the digital data, which is only a value, by defining it as voltage, the type of voltage (e.g., line-to-line voltage), unit (V, mV, kV), etc. The presentation layer will then code this information into the appropriate language used, e.g., ASCII format. The session layer will initiate the transmission of this information onto the network, and the transport layer will control the transmission, ensuring reliability of the information and security of the transmission.

The network layer adds the routing and address information. The data link layer provides the synchronization and error control bits, and the physical layer ensures that the complete frame at this stage is converted into the physical medium required for transmission, e.g., light pulses for a fiber optic cable.

This information, which consists of a frame of digital bits, is now transmitted over the physical transmission media. The receiver receives this data and, in a process reversed to the one above, the 'superficial' information is stripped away until only the application data remain.

The critical importance that the sender and receiver process the data in exactly the same way can clearly be seen, otherwise the receiver will incorrectly interpret the information, with possible disastrous consequences (easily imagined!), or not even receive the information at all.

EPA [It will be noted that the full seven-layer stack is actually only required for network communications. For point-to-point type connections or for more basic networks using a communication technique where a master or bus administrator is present, the full seven stacks are not required. The enhanced protocol architecture (EPA) has been developed for this purpose, using a three-layer stack. The three layers basically correspond to layers 1, 2 and 7 of the seven-layer model (i.e., physical, data link and application layers). Some of the features of the presentation layer still required will then be added to the application layer, and network features required will be added to the data link layer.]

This specific process by which the layers are added to the raw data actually defines the communication protocol. There are literally thousands of different ways of doing this, and thousands of different protocols have been used throughout the world at some stage or other. Initially protocols have not been clearly defined according to the OSI stack, and were developed according to the requirements of the specific manufacturer's equipment. Proprietary protocols were at the order of the day, and different manufacturers' equipment could as a rule not communicate to one another. Fortunately, certain protocols became widely accepted, and some manufacturers standardized on specific universally used protocols, e.g., Modbus, DNP3.0, etc.

Unfortunately the term 'protocol' is not defined very clearly. CSMA/CD, which is actually an access control principle has been termed a protocol in some literature, whereas it can only be seen as one part of a protocol associated with the network, transport and session layers.

Similarly, Ethernet is commonly called a protocol, but it only defines the physical and data link layers, and has subsequently been termed an 'uncomplete protocol' by various literature and presentations. Following this line of terminology, sometimes one protocol defines all seven (or the three required) layers of the OSI stack, and sometimes a collection of protocols, on top of each other, is required to form a 'complete' protocol.

Some of the more popular and widely used communication protocols are listed in Chapter 7, with specific emphasis on protocols used in power system automation.

6.6 Performance criteria

6.6.1 Transmission speed

Before discussing the factors that affect transmission speed, it is necessary to clarify the terminology.

The 'signalling rate' of a communication link is a measure of how many times the physical signal changes per second and is expressed as the baud rate. If each change represents the value of one bit, the baud rate is equal to the bit rate, which is expressed in bits per second (bps), or multiples such as kbps, Mbps and Gbps (kilo, mega and gigabits per second). An example is an EIA-232 link in which a bit is read as either a logical 0 or a logical 1; in this case 2400 baud equals 2400 bps.

There are sophisticated modulation techniques, used particularly in modems, which allow more than one bit to be encoded within a signal change. The CCITT V.22bis full duplex standard, for example, defines a technique called quadrature amplitude modulation that effectively increases a baud rate of 600 to a data rate of 2400 bps. Irrespective of the methods used, the maximum data rate is always limited by the bandwidth of the link.

Although there is a tendency for the terms baud and bps to be used interchangeably, they may not always be the same. As the important characteristic of a link is its bit rate, speeds will be expressed in bps in this text.

6.6.2 Bandwidth

The single most important factor that limits communication speeds is the bandwidth of the link. Bandwidth is generally expressed in hertz (Hz), meaning cycles per second. This represents the maximum frequency at which signal changes can be handled before attenuation degrades the message. Bandwidth is closely related to the transmission medium, ranging from around 3000 Hz for the public telephone system to the GHz range for optical fiber cable.

As a signal tends to attenuate over distance, communications links may require repeaters placed at intervals along the link, to boost the signal level.

6.6.3 Signal-to-noise ratio

The signal-to-noise (S/N) ratio of a communications link is another important limiting factor. Sources of noise may be external or internal.

The maximum practical data transfer rate for a link is mathematically related to the bandwidth, S/N ratio and the number of levels encoded in each signalling element. As the S/N decreases, so does the bit rate.

6.6.4 Data throughput

As data is always carried within a protocol envelope, ranging from a character frame to sophisticated message schemes, the read data transfer rate will be less than the bit rate. The amount of redundant data around a message packet increases as it passes down the protocol stack in a network. This means that the ratio of non-message data to 'real' information may be a significant factor in determining the effective transmission rate, sometimes referred to as the throughput.

6.6.5 Error rate

Error rate is related to factors such as S/N ratio, noise and interference. There is generally a compromise between transmission speed and the allowable error rate, depending on the type of application. Ordinarily, an industrial control system has very little error tolerance and is designed for maximum reliability of data transmission. This means that an industrial system will be comparatively slow in data transmission terms. As data transmission rates increase, there is a point at which the number of errors becomes excessive. Protocols handle this by requesting a retransmission of packets. Obviously the number of retransmissions will eventually reach the point at which a higher apparent data rate actually gives a lower real message rate, because much of the time is being used for retransmission.

6.6.6 Response time

Response time can be defined as the period it takes from the instant a command or request is initiated from one station to another station until the instant the receiving station respond to that command or request.

The response time depends on the transmission speed as well as the media access method. When, for example, a cyclic polling method is used for media access, the response time can be relatively slow although the transmission speed may be fast, as each slave station has to wait for its turn.

Response times are of particular importance in power system automation, as discussed in Chapter 10.

7
Communication protocols

7.1 Overview

Some of the more popular and widely used communication protocols are listed in Table 7.1, with specific reference to protocols used in substations at present.

Protocol	Originally used by	Speed	Access principle	OSI layers
MODBUS	Gould-Modicon	19.2 kbps	Cyclic polling	1,2,7
SPABUS	ABB (exclusively)	19.2 kbps	Cyclic polling	1,2,7
DNP3.0	GE Harris	19.2 kbps	Cyclic polling (+)	1,2,7 (+)
IEC 60870-5	All	19.2 kbps	Cyclic polling	1,2,7
MODBUS +	Gould-Modicon		Token	1,2,7
PROFIBUS	Siemens	12 Mbps	Token	1,2,7
MVB	ABB	1.5 Mbps	TDM	1,2,7 (+)
FIP	Merlin-Gerin	2.5 Mbps	TDM	1,2,7
Ethernet + TCP/IP	All	10 Mbps	CSMA/CD	1–7
LON	ABB (exclusively)	1.25 Mbps	PCSMA/CD	1–7
UCA 2.0	GE	10 Mbps	CSMA/CD	1–7

Table 7.1
Protocols used in substations

7.2 Distributed network protocol (DNP V3.0)

7.2.1 Introduction

The distributed network protocol is a data acquisition protocol used mostly in the electrical and utility industries. It is designed as an open, interoperable, and simple protocol specifically for SCADA controls systems. It uses the master/slave polling method to send and receive information, but also employs sub-masters within the same system. The physical layer is generally designed around RS-232 (V.24), but it also supports other physical standards such as RS-422, RS-485, and even fiber optic. There is a large support within the SCADA industry to use DNP as the universal *de facto* standard for data acquisition and control.

7.2.2 Interoperability

The distributed network protocol is an interoperable protocol designed specifically for the electric utilities, oil, gas, and water/waste water and security industries. As a data acquisition protocol, the need to interface with many vendors' equipment was and is necessary. By having a certification process, the protocol ensures that different manufacturers are able to build equipment to the DNP standard. This protects the end-user when purchasing a certified DNP device. As more and more manufacturers produce DNP certified equipment, the choices and confidence of users will increase.

7.2.3 Open standard

The DNP was created with the philosophy of being a completely open standard. Since no one company owns the DNP standard it means that producers of equipment feel that they have a level playing field on which to compete. This allows different manufactures to have equal input into changes to the protocol. In addition, it means that the cost to develop a system is reduced. The producer does not need to design all parts of the SCADA system. In a proprietary system, the manufacture usually has to design and produce all parts of the SCADA system, although some of those parts may not be so profitable. One manufacturer is free then to specialize on a few products that is its core business.

7.2.4 IEC and IEEE

DNP is based on the standards of the International Electrotechnical Commission (IEC) Technical Committee 57, Working Group 03 who have been working on an OSI 3 layer 'Enhanced Performance Architecture' (EPA) protocol standard for telecontrol applications. DNP has been designed to be as close to compliant as possible to the standards as they existed at time of development with the addition of functionality not identified in Europe but needed for current and future North American applications. Recently DNP 3.0 was selected as a Recommended Practice by the IEEE C.2 task force; remote terminal unit to intelligent end device's communications protocol.

7.2.5 SCADA

The DNP is well developed as a device protocol within a complete SCADA system. It is designed as a data acquisition protocol with smart devices in mind. These devices can be coupled as a multidrop fieldbus system. The fieldbus DNP devices are integrated into a software package to become a SCADA system. DNP does not specify a single physical layer for the serial bus (multimode) topology. Devices can be connected by 422 (four wire), 485 (two wire), modem (Bell 202) or with fiber optic cable. The application program can integrate DNP with other protocols if the SCADA software permits. Using tunneling or encapsulation the DNP could be connected to an intranet or the Internet.

7.2.6 Development

The specification was first developed by the GE Harris Company but has been released under the DNP Users Group since 1992. Now over 100 vendors offer DNP V3.0 products. These products range from master stations to intelligent end-devices. The protocol is designed so that a manufacturer can develop a product that supports some but not all of the functions and services that DNP supports. The DNP 3.0 was derived from an earlier version of the IEC 870.5 specs. The DNP Users Group now controls the documenting and updating of the protocol. For users of the DNP a copy of the protocol can be purchased through the website http://www.dnp.org

7.2.7 Physical layer

The physical layer of DNP is a serial bit oriented asynchronous system using 8 data bits, 1 start bit, 1 stop bit and no parity. Synchronous or asynchronous is also allowed. It has two physical modes of operation, direct mode (point-to-point) or serial bus mode (multidrop). The two modes are not usable at the same time. Both modes can be half- or full-duplex. With either mode, a carrier detection system must be used. The DNP protocol is a modified master/slave system. Multi-masters are allowed but only one device can be a master at a time. There are possible collisions on the system. The configuration of the physical layer determines the method of collision avoidance or recovery. The DNP can prioritize devices in a multi-master mode.

7.2.7.1 Physical topologies

The DNP protocol supports five communication modes: two-wire point-to-point, two-wire multidrop, four-wire point-to-point, four-wire multidrop, and dial-up modems. A system with only two nodes, a master, and a slave is called a direct bus. If the system is multidrop with multiple nodes, it is called a serial bus. Both systems can use two- or four-wire connection methods. The two-wire method can only run half-duplex while the four-wire method can run either half- or full-duplex. The DNP supports multiple master, multiple slave, and peer-to-peer communications.

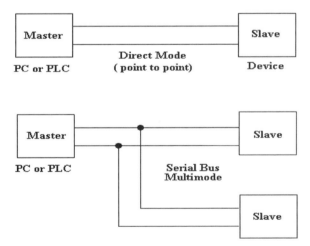

Figure 7.1
Direct and serial modes

7.2.7.2 Modes

Two-wire point-to-point

The DNP protocols physical layer supports point-to-point communications. The two-wire half-duplex mode usually uses RS-485 or a two-wire modem as a physical system. If a modem is used then the interface to the modem usually uses the V.24 ITU standard (RS-232). The two-wire mode does not support a full-duplex operation within DNP, only half-duplex. With the point-to-point mode there is no possibility of collisions. The master transmits the frame and then the slave responds. The only concern is the time it takes for a response due to propagation delays. There is a configuration in DNP for setting up this time. (master_min time)

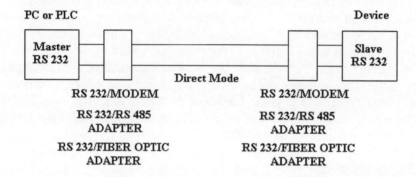

Figure 7.2
Two-wire direct mode

Two-wire multidrop

The DNP physical layer supports multi-point communications. The two-wire multidrop mode usually uses RS-485, fiber optic or Bell 202 modems as a physical system. The two-wire mode does not support a full-duplex operation within DNP, only half-duplex. With the multidrop two-wire mode, there are possibilities of collisions. This is possible because two masters or slaves can access the line at the same time. To overcome this DNP inserts a time-delay after the loss of carrier. Carrier is an indication that someone is transmitting on the two-wire bus. In a two-wire multimode, all devices on the line must have some way of determining that someone is transmitting.

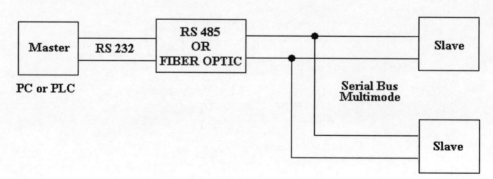

Figure 7.3
Two-wire multimode

Four-wire point-to-point

Four-wire point-to-point is used within DNP as a full-duplex master-to-slave system. The physical standards used are RS-422, RS-232, and four-wire modems. Since this mode is only point-to-point there is no problem with collisions. However, V.24 is used as a handshaking system to control the communications. This includes DCD (data carrier detect). Propagation delays are also used to allow the devices time to detect the loss of carrier. Four-wire communications allow for true full-duplex communication, but in practice, it is rarely implemented.

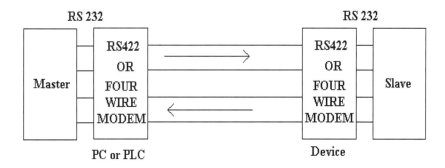

Figure 7.4
Four-wire direct mode

Four-wire multi-point

DNP allows for a four-wire multi-point mode. This mode can use half- or full-duplex communications, although again full-duplex is rarely used. The reason full-duplex is not

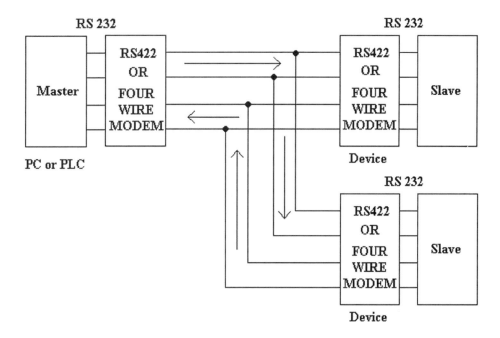

Figure 7.5
Four-wire serial mode

used is because of the complexity of collision avoidance. Have multiple devices all talking at the same time in both directions is difficult at best to implement. The master does not have a problem with collisions but may have problems with other primary masters accessing the line before the slave has a chance to respond. The slave can have many problems with collisions because many of them may want to respond to their masters at the same time. One way that DNP uses to handle these collisions is to allow the slaves to collide. This causes the second master to time out and the primary master to gain access to the bus. The master can then send out high priority messages.

Dial-up modem

The DNP supports the use of a dial-up modem mode. This mode is a point-to-point circuit. It usually uses V.24 as a connection system (RS-232). DCD is used differently in this case because carrier detect in modems means that a line has been established, not that data is being sent. The RTS line is placed high to tell the modem that the DTE wishes to send data. The CTS is placed high by the modem to tell the DTE it is OK to send data. There is no way for the local end to tell the remote end that data is being transferred. It is then up to the remote end to be able to detect data coming in.

Figure 7.6
Dial-up modem mode

7.2.8   Data link layer

The data link layer of the DNP defines the frame size, shape, length, and contents. DNP uses the convention of the octet instead of byte. The DNP uses hexadecimal as a language within the frame. The frame is laid out as follows:

7.2.8.1   Frame outline

Start (2 octets) 0X0564   (0000010100100000)
Length (1 octet) 5 to 255   (decimal)
Control (1 octet) includes function code
Destination (2 octets)
Source (2 octets)
CRC (2 octets) for the length, control, destination, and source
User data (16 octets)
CRC (2 octets) for the user data above
More user data (16 octets)
CRC (2 octets) for the user data above only
More user data (1 to 16 octets) variable
CRC (2 octets)
END

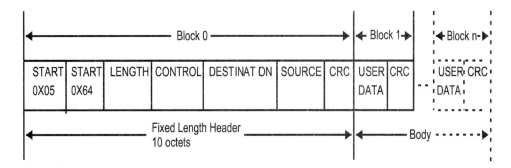

Figure 7.7
DNP packet format

Note that the user data can go on until the maximum number of octets is reached. This is determined by the length octet. The maximum octets of data are 255 and the minimum is 5. The last user data may have less than 16 octets. Each CRC is calculated for that user data, not for the whole frame.

7.2.8.2 Function codes

Control octet uses four bits to determine how the data link will handle the frame. There are six basic function codes:

- Reset – This function code is used to synchronize a primary and secondary station for further send-confirm transactions.
- Reset of user process – This function code is used to reset the data link user process.
- Test – The test command is used to test the state of the secondary data link.
- User data – The user data function is used to send confirmed data to a secondary station.
- Unconfirmed user data – This function is used to send user data to the secondary station without needing confirmation.
- Request link status – This command is used to request the status of the secondary data link.

7.2.8.3 Examples of transmission procedures

Reset of secondary link

In the figure below, a primary station sends a send-confirm reset frame to a secondary station. The secondary station receives the message and responds with an ACK confirm frame.

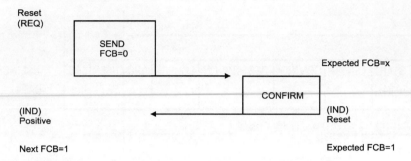

Figure 7.8
Reset of secondary link

Reset of user process

In the figure below, a primary station sends a send-confirm reset user process frame to a secondary station. The secondary station receives the message and responds with an ACK confirm frame.

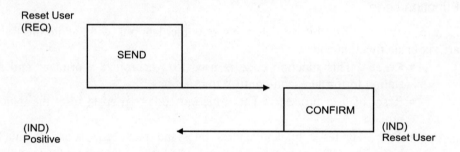

Figure 7.9
Reset of user process

Send-confirm user data

In the figure below, the designated master station acting as a primary station sends a send-confirm frame to a non-master station acting as a secondary station. This is the first frame with PCV valid after the link was reset so FCB = 1 in the send frame. The secondary station n expects FCB to be 1 since this is the first frame after the link was reset and sends a confirm frame. The master station upon receiving the confirm frame assumes the message was correctly received and indicates success to the master station data link user.

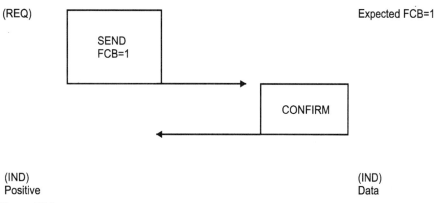

Figure 7.10
Send-confirm user data

Send-No reply expected

In the figure below, the master or non-master primary station sends 3 frames to the secondary master or non-master. Upon successfully transmitting the send frame, the primary station indicates success to the data link user. The secondary station, upon reception of a valid frame indicates data availability to the data link user.

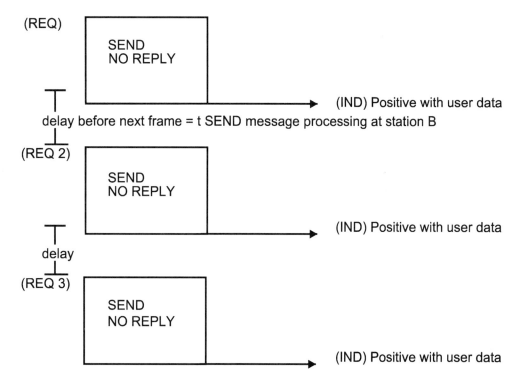

Figure 7.11
Send-no reply expected

Send-NACK

In the figure below, a non-master primary station sends a frame to the master secondary. Upon reception of the first confirm, the primary indicates success to the data link user. The primary sends a second frame to the secondary. The secondary master decides that it cannot accept any frames at this time and sends a NACK frame back. The primary, after receiving this NACK, will fail the transaction and send a negative indication to the data link user.

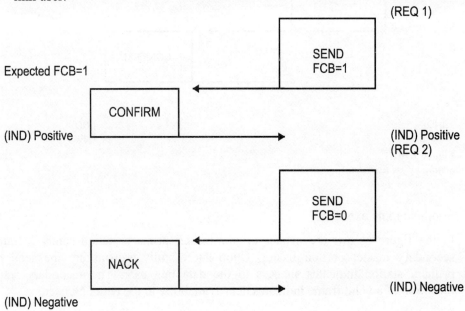

Figure 7.12
Send/NACK

Request-respond

In the figure below, a primary station sends consecutive frames to a secondary station. When the secondary station cannot receive any more frames, the confirm message contains the DFC bit set. The primary station will, upon receiving the confirm, stop sending data frames to the secondary station but will instead periodically request the status of the secondary by sending a request-respond frame. The secondary will respond to the request frame with the current state of the DFC. If the secondary is ready to receive more data, the DFC returned will be 0 otherwise the DFC returned will be 1. When the primary station recognizes DFC = 0 in the respond frame, the transmission of send frames will continue.

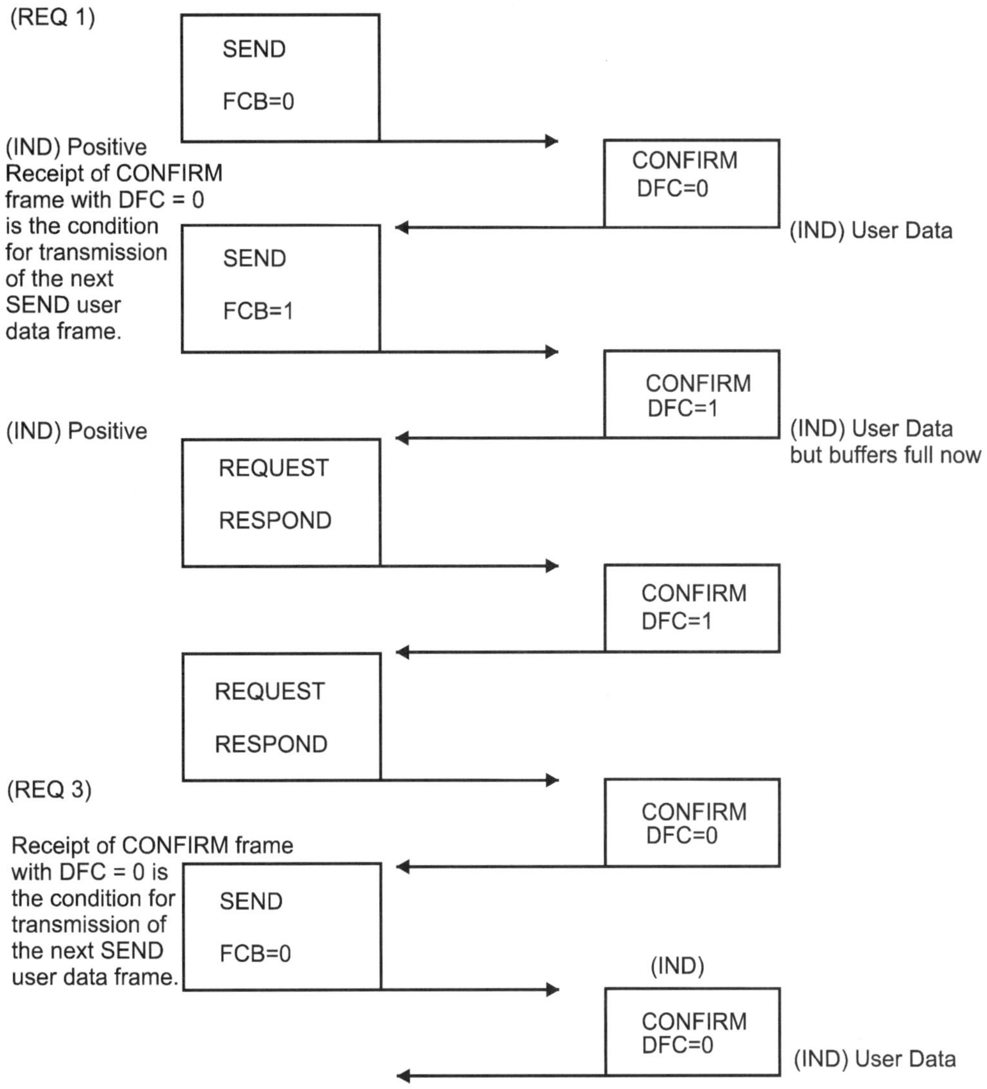

Figure 7.13
Request-respond

7.2.9 Transport layer (pseudo-transport)

The distributed network protocol does not support a true transport layer as defined by the ISO open system interconnection model. It does support a pseudo-transport layer known as the super-data link transport protocol. This is because some of the functions of the data link layer do not strictly meet the ISO OSI model. These functions are then moved out of the data link layer and placed in this pseudo-transport layer. These data link functions consist of breaking the transport service data unit (TSDU) into smaller sequenced frames called link service data units (LSDU). Each of these frames has transport protocol control information. The maximum size of an LSDU is 249 octets. This is done to reduce the length of packets in case of errors. If a packet is in error, then a retry will be initiated. A shorter packet means that the retries will be quicker.

7.2.10 Application layer

The DNP supports an application layer by defining an extensive data object library, function codes and message formats for both the requestor and the response devices. These are used in the user layer to build a final application. Once this application is built and the data objects, function codes, and message formats are absorbed the application becomes the application layer. A complete list of data objects and function codes can be found in the DNP Version 3.0 standard document available from DNP Users Group http://www.dnp.org. And other information can be found at Harris Controls Division. http://www.harris.com/harris/search.html

7.2.11 Conclusion

The distributed network protocol only supports the physical layer, data link layer, and application layers within the open system interconnection model. The physical layer is the least supported. DNP is based on the enhanced protocol architecture (EPA), a protocol standard for telecontrol applications. It supports advanced RTU functions and messages larger than the normal frame length. It takes user data and breaks it up into several sequenced transport protocol data units (TPDU) each with transport protocol control information (TPCI). The TDPU is sent to the data link layer as a link service data unit. The receiver receives multiple TPDUs from the data link layer and assembles them into one TSDU.

There is no official compliance testing, but there is help online. If a vendor claims to comply with one of the DNP V3.00 sub-set definitions, then the device is definitely interoperable. Of course interoperable does not mean efficient. It is often best to stick with one supplier when possible. The DNP is truly an open non-proprietary interoperable protocol.

7.3 Modbus

7.3.1 Overview

Modbus was originally developed by Gould-Modicon as a proprietary protocol for PLC communication. It is now an 'open' protocol used by thousands of vendors. It is mainly used in local serial connections and is not suited to be used as a telemetry protocol.

Modbus is a simple master–slave protocol, using the cyclic polling method. It applies to layers 1, 2 and 7 of the OSI stack.

In contrast to the many other buses discussed, no interface is defined in the modbus protocol. The user can therefore choose between EIA-232, EIA-422, EIA-485 or 20 mA current loop, all of which are suitable for the transmission rates that the protocol defines.

Although the modbus is relatively slow in comparison with other buses, it has the advantage of wide acceptance among instrument manufacturers and users. About 20 to 30 manufacturers produce equipment with the modbus protocol and many systems are in industrial operation. It can therefore be regarded as a *de facto* industrial standard with proven capabilities. A recent survey in the well-known *American Control Engineering* magazine indicated that over 40% of industrial communication applications use the modbus protocol for interfacing.

However, modbus is severely limited as a protocol for electrical substation communications.

The modbus is accessed on the master/slave principle, the protocol providing for one master and up to 247 slaves. Only the master initiates a transaction.

Transactions are either a query/response type where only a single slave is addressed, or a broadcast/no response type where all slaves are addressed. A transaction comprises a single query and single response frame or a single broadcast frame.

Certain characteristics of the modbus protocol are fixed, such as frame format, frame sequences, handling of communications errors and exception conditions and the functions performed. Other characteristics are selectable. These are transmission medium, transmission characteristics, and transmission mode, RTU or ASCII. The user characteristics are set at each device and cannot be changed when the system is running.

The modbus protocol provides frames for the transmission of messages between master and slaves. The information in the message is the address of the intended receiver, what the receiver must do, the data needed to perform the action and a means of checking errors. The slave reads the messages, and if there is no error it performs the task and sends a response back to the master. The information in the response message is the slave address, the action performed, the result of the action and a means of checking errors. If the initial message was of broadcast type, there is no response from the slaves.

Normally, the master can send another query as soon as it has received the response message. A timeout function ensures that the system still functions when the query is not received correctly.

Data can be exchanged in two transmission modes:

- ASCII – readable, used e.g., for testing
- RTU – compact and faster; used for normal operation (Hex)

The RTU mode (sometime also referred to as modbus-B for modbus binary) is the preferred modbus mode and will be discussed in this section. The ASCII transmission mode has a typical message that is about twice the length of the equivalent RTU message.

The modbus also provides an error check for transmission and communication errors. Communication errors are detected by character framing, a parity check and a redundancy check or CRC. The latter varies depending on whether the RTU or ASCII transmission mode is being used.

7.3.2 Modbus functions

All functions supported by the modbus protocol are identified by an index number. They are designed as control commands for field instrumentation and actuators and are as follows:

- Coil control commands for reading and setting a single coil or a group of coils
- Input control commands for reading input status of a group of inputs
- Register control commands for reading and setting one or more holding registers
- Diagnostics test and report functions
- Program functions
- Polling control functions
- Reset

7.4 Modbus plus

7.4.1 Introduction to modbus plus network

Modbus plus is a local area network (LAN) system, which is exclusively designed for industrial control applications. Its main characteristics are:

- Supports up to 64 addressable node devices.
- Host level peer-to-peer communication.
- Distributed input/output (DIO) communications.
- Direct connection of 32 devices with a bus cable, up to 450 m long, and if extended with a repeater up to 64 devices, with a total cable length of 1800 m.
- Host level networks can be joined through bridge plus devices.

7.4.2 Modbus network terminology

Node – Any device that is physically connected to the modbus plus cable.
Cable segment – A single length of trunk cable between passive devices that provide connections for the trunk, called taps.
Section – A series of nodes, joined only by cable segments.
Token – A grouping of bits, passed from one device to another on a single network, to grant access for sending messages.
Network – The grouping of nodes on a common signal path that is accessed by a passing of a token.

7.4.3 How the network works

Each network node is identified by an address, assigned by the user. Addresses are within the range of 1 to 64 decimal and do not have to be sequential.

The lowest address node issues a token and starts it rotating in a fixed sequence from one node to the next. The token cannot be passed to another network through a bridge. Each network has its own token.

When it possesses a token a node can transmit messages to any other node. Each message contains information about its source and destination, including its routing path through bridges, if it has to reach a remote network node.

There is a field containing global data within a token. Before the node passes the token it can place data in this field to update the global database within the network when the token is placed back on the bus. Other nodes monitor the token pass, and may extract the global data. This allows rapid updating of alarms, setpoints, and other data.

7.4.4 Overview of the physical network

The network bus consists of twisted pair shielded cable. The two data lines in the cable are not sensitive to polarity. There is a wiring standard to facilitate maintenance.

Each network cable section can support up to 32 nodes, with a maximum cable distance of 450 m. Sections can be joined by repeaters to extend the cable length up to 1800 m and to increase the number of nodes up to 64. The cable length between any pair of nodes has to be a minimum of 3 m and a maximum of 450 m.

On dual-cable networks the cables are known as cable A and cable B. Each section can be up to 450 m. The difference in length between cables A and B must not exceed 150 m, between any pair of nodes.

Nodes are connected to the cable by means of a tap device, supplied by Modicon. This provides 'through' connections for the network trunk cable, 'drop' connections to the node device and a grounding terminal.

The tap also contains a resistive termination that is connected by two internal jumpers. If a tap is installed at the end of a section, these jumpers have to be connected to prevent signal reflections. If a tap is installed inline, the jumpers have to be removed.

7.4.5 How nodes access the network

Nodes layout – When a network is switched on each node becomes aware of the other active nodes. Initial ownership of the token is established and its rotation begins. It is your choice to lay the network out as one large one, or as several smaller networks, connected through bridges. As the token is not passed through bridges it is possible in that way to vary the timing of the complete token rotation. For example, fast token rotation in a small network can deal with high-priority data transfer, (which may be critical), while lower-priority data to outer nodes can still be passed through bridges.

Token rotation sequence – It is determined by the node addresses. Token rotation begins at the lowest-addressed node and proceeds consecutively, until it reaches the highest-addressed node. Then it is transferred again to the lowest-addressed node and the process repeats.

Point-to-point message transactions – While a node holds a token it is permitted to send its application messages. Simultaneously the other nodes monitor the network for incoming messages. When a node receives a message it sends an immediate acknowledgment to the originating node. If in this message the receiving node is requested to send some data, it will send it back, but only when it will be in possession of the next token.

The only information, which a node is permitted to transmit without holding the token, is statistics. Such information includes identification of active nodes, current software version, network activity, and error reporting. If a node is requested for such information, it can be embedded in the acknowledgment, and sent back immediately.

Global database transactions – When a node passes the token it can transmit up to 32 words (16 bits each) of global information to all other nodes in the network. Although only one node accepts the token pass, all nodes monitor the token transmission and read its content. Each one of them maintains a table of global data. Then according to its specific tasks it determines when and how to use it. Global database applications include time synchronization, rapid notification of alarm conditions, and multicasting setpoint values to all devices in a common process.

7.4.6 Error checking and recovery

The program senses various deviations from normal operation. For example, if no acknowledgment is received after a message transmission, the node will re-attempt to send the message again. After three unsuccessful attempts, the node will set an error, which can be sensed by the application program.

If two nodes have been assigned the same address, the application program will ensure that the second one remains silent, will handle the rest of the application and will set an error, which can be sensed by the application program.

If after passing the token a node does not sense any valid activity from its successor, it will try to pass the token again. After the second unsuccessful attempt the faulty node will remain silent and a new token sequence will be established.

7.4.7 Peer cop transactions

This is a method of peer-to-peer communication between nodes, in which data is transferred as part of the passing token. Nodes have to be specifically configured to send and receive data, otherwise they will ignore it. There are four kinds of peer cop (peer-cop?) transactions:

- Global output – Each node can send up to 32 words of data globally to all other nodes.
- Specific output – Each node could send up to 32 words of data specifically to another node, which could be situated anywhere in the network. Multiple node destinations have to be specified with a maximum of 500 data words.
- Global input – Each node could receive up to 32 words of global data from any other node.
- Specific input – Each node could receive up to 32 words of data from any specific node in the network.

The net effect of using peer cop is that each sending node can specify unique references as data sources. On the other hand each receiving node can specify the same as data destinations. Thus when receiving global data each node can refer only to these specific locations in incoming data, which are of interest to it. In this manner data is transferred rapidly as part of each token pass and can be directly mapped in sending and receiving nodes.

7.5 LonTalk

LonTalk is part of the LonWorks technology developed by Echelon Corporation. LonTalk is the protocol, whereas LonWorks comprises the entire range of hardware, software, transceivers, network management, program interfaces, installation and configuration tools, etc.

The LonTalk protocol implements all seven layers of the ISO/OSI reference model. The protocol is embedded as a mixture of hardware and firmware on a silicon chip, manufactured worldwide by Motorola and Toshiba. The protocol supports peer-to-peer communication, and includes features such as transaction acknowledgment, sender authentication, priority transmissions, duplicate message detection, collision avoidance, unicast/multicast/broadcast addressing, error detection, and recovery.

Table 7.2 illustrates the mapping of LonTalk services onto the seven-layer OSI model:

	OSI Layer	Purpose	Services Provided	Processor
7	Application	Application compatibility	Standard network variable types	Application
6	Presentation	Data interpretation	Network variables, frame transmission	Network
5	Session	Remote actions	Request/response, authentication, network management	Network
4	Transport	End-to-end reliability	Acknowledged/Unacknowledged, Unicast/multicast, ordering, duplicate detection	Network
3	Network	Destination addressing	Addressing, Routing	Network
2	Link	Media access and framing	Data encoding, error checking, PCSMA, collision detection and avoidance, priority	MAC
1	Physical	Electrical interconnect	Media specific interfaces and modulation	MAC, XCVR

Table 7.2
LonTalk layers

The following are the main features of the LonTalk protocol:

7.5.1 Multiple media support

A wide variety of communication media is supported, including twisted pair, powerline, RF, infrared, coaxial cable, and fiber optics.

7.5.2 Multiple communication channels

A channel is a physical transport medium and can contain up to 32 385 nodes. A network may consist of two or more channels. Multiple channels and networks are supported by LonTalk, and network loading can be optimized by localizing traffic.

7.5.3 Communication rates

Different channels may be configured for different bit rates to allow for differences in distance, throughput and/or other network requirements. Bit rates vary in discrete steps from a minimum of 0.6 kbps to a maximum of 1.25 Mbps.

7.5.4 Addressing

The top level of the addressing hierarchy is the domain. Different domains may be specified if different network applications are implemented on a shared communications medium. A domain consists of up to 255 subnets.

The second level is the subnet. A subnet is a logical grouping of nodes from one or more channels. There may be up to 127 nodes per subnet.

The lowest level of addressing is the specific node. There may be up to a maximum of $255 \times 127 = 32\,385$ nodes in a single domain. One node may be a member of one or two domains, allowing a node to serve as an inter-domain gateway.

Nodes may also be grouped. Up to 256 groups may be specified in a domain, and up to 64 nodes may be in a group for acknowledged service. An unlimited number of nodes may belong to a group for unacknowledged service. A specific node may be a member of up to 15 groups for receiving messages. Group addressing allows many nodes to receive information using a single message on the network, and reduces the length of the addressing data send with each message.

The communication channel does not affect the addressing of the node. Domains can contain several channels and subnets and groups may also span several channels.

Each node has a unique ID assigned during manufacture. This ID is normally used as a network address only during installation and configuration.

Nodes are addressed in the following format:

Address data specified	Nodes addressed
Domain, Subnet = 0	All nodes in the domain
Domain, Subnet	All nodes in the subnet
Domain, Subnet, Node	Specific logical node
Domain, Group	All nodes in the group
Unique ID	Specific physical node

7.6 Ethernet

7.6.1 Background

The Ethernet network concept was developed by Xerox Corporation at its Palo Alto Research Center (PARC) in the mid seventies. It was based on the work done by researchers at the University of Hawaii where there were campus sites on the various islands. Its ALOHA network was setup using radio broadcasts to connect the various sites. This was colloquially known as their 'Ethernet' since it used the 'ether' as the transmission medium and created a network 'net' between the sites. The philosophy was quite straightforward. Any station that wanted to broadcast to another station would do so immediately. The receiving stations then had a responsibility to acknowledge the message; thus advising the original transmitting station of a successful reception of the original message. This primitive system did not rely on any detection of collisions (two radio stations transmitting at the same time) but rather waited for an acknowledgment back within a predefined time.

The initial system installed by Xerox was so successful that they soon applied the system to their other sites typically connecting office equipment to shared resources such as printers and large computers acting as repositories of large databases, for example.

In 1980, the Ethernet Consortium consisting of Xerox, Digital Equipment Corporation and Intel (sometimes called the DIX consortium) issued a joint specification based on the Ethernet concepts and known as the Ethernet Blue Book 1 specification. This was later superseded by the Ethernet Blue Book 2 specification, which was offered to the IEEE as a standard. In 1983, the IEEE issued the 802-3 standard for carrier sense; multiple access; collision detect (CSMA/CD) LANs based on the Ethernet standard which gave this networking standard even more credibility.

As a result of this, there are three standards in existence. The first – often termed Ethernet Version 1 – can be disregarded as very little equipment based on this standard is still in use. Ethernet Version 2 or 'Blue Book Ethernet' is, however, still in use and there is a potential for incompatibility with the IEEE 802.3 standard. Despite the generic term 'Ethernet' being applied to all CSMA/CD networks, it should actually be reserved for the original DIX standard. This manual will continue with popular use and refer to all the LANs of this type as Ethernet, unless it is important to distinguish between them.

Ethernet uses the CSMA/CD access method discussed in Chapter 6. This gives a system that can operate with little delay, if lightly loaded, but the access mechanism can fail completely if too heavily loaded. Ethernet is widely used commercially, and the end equipment is relatively cheap and produced in vast quantities. Because of its probabilistic access mechanism, there is no guarantee of message transfer and messages cannot be prioritized. It is becoming more widely used industrially despite these disadvantages.

The Ethernet standard only defines the physical layer and data link layer of the ISO/OSI stack. The IEEE 802.3 standard defines a range of cable types that can be used for a network based on this standard. They include coaxial cable, twisted pair cable, and fiber optic cable. In addition, there are different signaling standards and transmission speeds that can be utilized.

7.6.2 Signaling method

Ethernet signals are encoded using the Manchester encoding scheme, as illustrated in Figure 7.14. This method allows a clock to be extracted at the receiver end synchronize the transmission/reception process. The encoding is performed by an EXCLUSIVE-OR between a 20 MHz-clock signal and the data stream. In the resulting signal, a '0' is represented by a high to low change at the center of the bit cell, whilst a '1' is represented by a low to high change at the center of the bit cell. There may or may not be transitions at the beginning of a cell as well, but these are ignored at the receiver. The transitions in every cell allow the clock to be extracted, and synchronized with the transmitter.

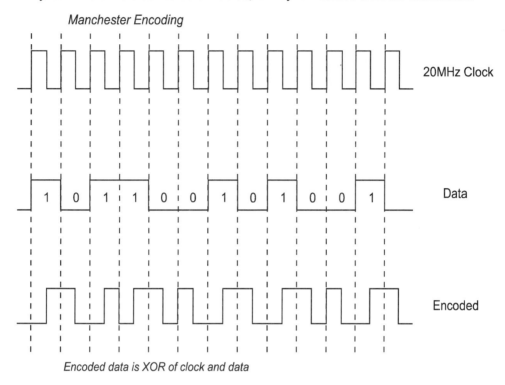

Figure 7.14
Manchester encoding

The voltage swings were from –0.225 to –1.825 volts in the original Ethernet specification. In the 802.3 standard, voltages are specified to oscillate between 0 and –2.05 volts with a rise and fall time of 25 ns at 10 Mbps.

7.6.3 Medium access control

Essentially, the method used is one of contention. Each node has a connection via a transceiver to the common bus. As a transceiver, it can both transmit and receive at the same time. Each node can be in any one of three states at any time. These states are:
- idle, or listen
- transmit
- contention

In the idle state, the node merely listens to the bus, monitoring all traffic that passes. If a node then wishes to transmit information, it will defer whilst there is any activity on the bus, since this is the 'carrier sense' component of the architecture. At some stage, the bus will become silent, and the node, sensing this, will then commence its transmission. It is now in the transmit mode, and will both transmit and listen at the same time. This is because there is no guarantee that another node at some other point on the bus has not also started transmitting having recognized the absence of traffic.

After a short delay as the two signals propagate toward each other on the cable, there will be a collision of signals. Quite obviously, the two transmissions cannot coexist on the common bus, since there is no mechanism for the mixed analog signals to be 'unscrambled'. The transceiver quickly detects this collision, since it is monitoring both its input and output and recognizes the difference. The node now goes into the third state of contention. The node will continue to transmit for a short time – the jam signal – to ensure the other transmitting node detects the contention, and then performs a back-off algorithm to determine when it should again attempt to transmit its waiting frames.

Collisions are a normal part of a CSMA/CD network. The monitoring and detection of collisions is the method by which a node ensures unique access to the shared medium. It is only a problem when there are excessive collisions. This reduces the available bandwidth of the cable and slows the system down while retransmission attempts occur.

The principle of collision cause and detection is shown in Figure 7.15.

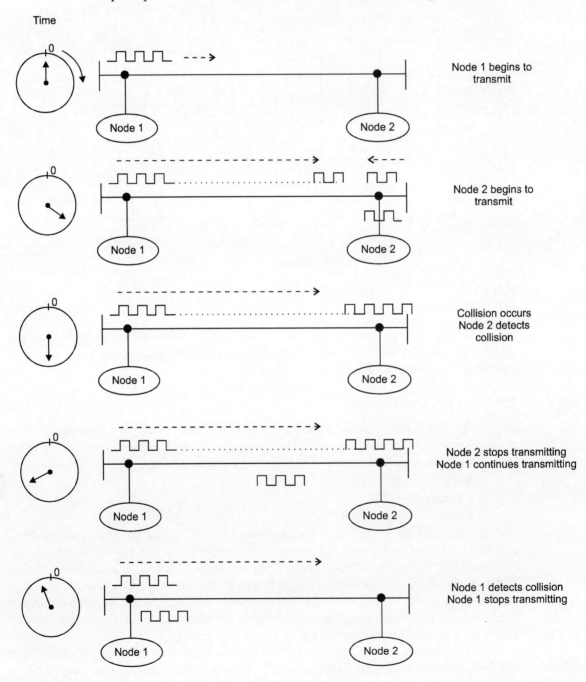

Figure 7.15
CSMA/CD collisions

7.7 TCP/IP

7.7.1 Origin and background of TCP and IP

The Internet was originally known as the Advanced Research Projects Agency Network (ARPANET)) and was built by Bolt, Beranek, and Newman Inc. (BBN). This system operated from 1969 to 1990 and was the template, or design base for TCP/IP.

In the early 1960s, The American Department of Defense (DoD) indicated the need for a wide-area, cross platform communication system. To accommodate this the ARPA system was renamed the [United States] Defense Advanced Research Projects Agency (DARPA), and it used the Xerox Networking System (XNS) protocol, later known as Ethernet.

However, this particular protocol was found to be inadequate on its own, and as a result, the TCP/IP protocol suite was developed for use on top of the Ethernet, which only serves as the physical and data link layer. Comparing to the ISO/OSI stack, the transport layer is provided by TCP (transport control protocol) and the network layer by IP (Internet protocol).

In 1967 the Stanford Research Institute was contracted to develop this new suite of protocols, with the resulting timetable of development occurring:

1970: Commencement of the development
1972: Approx. 40 sites connected and TCP/IP support commenced
1973: The first international connection made
1974: TCP/IP released to the public

Initially TCP/IP was used to interconnect government, military and educational sites together, slowly connecting to commercial companies as time progressed.

7.7.2 The OSI and TCP/IP model

The open system interconnection – reference model (OSI-RM) is a suggested reference model, which isolates various parts of the networking layers from all other layers other than the two layers directly adjacent ('below' and 'above') to it. It was designed to allow internetworking between dissimilar computing systems.

It is a development from the ARPA 4 level model that was used specifically for academic, government, and military systems in the U.S.A and was a very specific model, as illustrated in Figure 7.16.

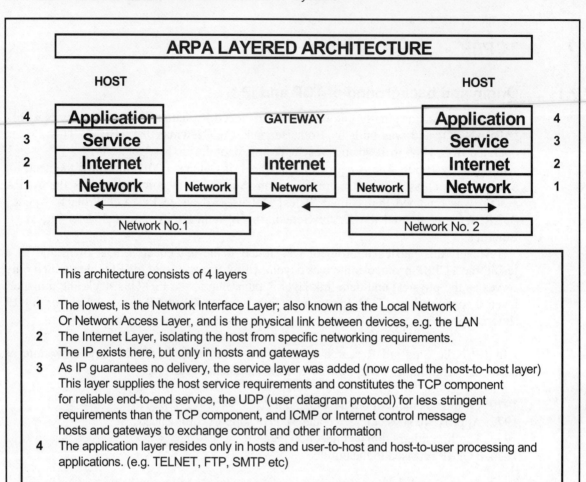

Figure 7.16
The ARPA architecture

Figure 7.16 illustrates the ARPA network with its four layers of application. Figure 7.17 below illustrates the contents and usage of each of the layers; against a common template that is also used later to illustrate the OSI layers.

ARPA LAYER	PROTOCOL IMPLEMENTATION					
PROCESS AND APPLICATION	File Transfer	Electronic Mail	Terminal Emulation	File Transfer	Client/Server	Network Management
	File Transfer Protocol (FTP)	Simple Mail Transfer Protocol (SMTP)	TELNET Protocol	Trivial File Transfer Protocol	Sun Microsystems. Network file Systems Protocol (NFS)	Simple Network Management Protocol (SNMP)
	MIL-STD 1780 RFC 959	MIL-STD 1781 RFC 821	MIL-STD 1782 RFC854	RFC 783	RFCs 1014, 1057 & 1094	RFC 1157
HOST - TO - HOST	Transmission Control Protocol (TCP) MIL-STD 1778 RFC 793			User Datagram Protocol (UDP) RFC 768		
INTERNET	Address Resolution ARP RFC 826 & RARP RFC 903		Internet Protocol (IP) MIL STD 1777 & RFC 791		Internet Control Message Protocol (ICMP) RFC 792	
NETWORK INTERFACE	Network Interface Cards: Ethernet, Token-Ring, ARCNET, MAN and WAN. RFC 894, 1042, 1201 and others					
	Transmission Media: Twisted pair cable, Coaxial Cable, Fiber Optics, Wireless Media etc. etc.					

Figure 7.17
The basic ARPA model

Commercial and public requirements have a much broader set of needs and uses and as such, the seven-layer OSI-RM model was developed in order to accommodate these.

Below, in Figure 7.18, is an example of the OSI-RM, from which it can be easily seen that (for example) the transport layer only has to communicate with the layers located either side of it, the network and session layers in this example. All the other layers are not directly linked to the transport layer, so that the transport layer has no direct impact on and receives no direct communications from them.

OSI LAYER	PROTOCOL IMPLEMENTATION					
APPLICATION	File Transfer	Electronic Mail	Terminal Emulation	File Transfer	Client/Server	Network Management
PRESENTATION	File Transfer Protocol (FTP)	Simple Mail Transfer Protocol (SMTP)	TELNET Protocol	Trivial File Transfer Protocol	Sun Microsystems. Network file Systems Protocol (NFS)	Simple Network Management Protocol (SNMP)
SESSION	MIL-STD 1780 RFC 959	MIL-STD 1781 RFC 821	MIL-STD 1782 RFC854	RFC 783	RFCs 1014, 1057 & 1094	RFC 1157
TRANSPORT	Transmission Control Protocol (TCP) MIL-STD 1778 RFC 793			User Datagram Protocol (UDP) RFC 768		
NETWORK	Address Resolution ARP RFC 826 & RARP RFC 903		Internet Protocol (IP) MIL STD 1777 & RFC 791		Internet Control Message Protocol (ICMP) RFC 792	
DATA LINK	Network Interface Cards: Ethernet, Token-Ring, ARCNET, MAN and WAN. RFC 894, 1042, 1201 and others					
PHYSICAL	Transmission Media: Twisted pair cable, Coaxial Cable, Fiber Optics, Wireless Media etc. etc.					

Figure 7.18
The basic OSI model

The OSI-RM is usually just referred to as the OSI model. Figure 7.19 illustrates the relationships between the OSI and ARPA models.

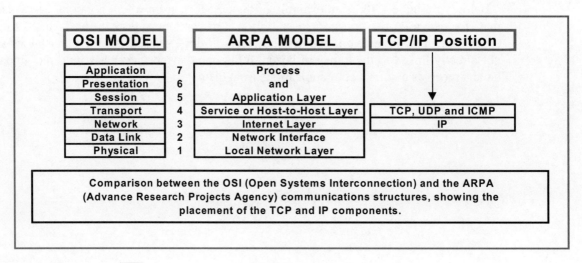

Figure 7.19
Comparison between the ARPA and OSI models

7.7.3 TCP/IP, OSI, and ISO standards

There are some sets of references to protocols that fall under the ISO standards for the interconnection of computing systems. Generically, TCP/IP is a type of OSI protocol; that is, it describes the structuring of the protocols involved. However, under the ISO standard TCP/IP is NOT an OSI protocol.

This looks like a contradiction of terms. However, if one remembers the meaning of the term 'protocol', that of being a set of rules governing the exchange or transmission of data between devices, it can be seen that the TCP/IP and OSI protocol models perform very similar functions.

As such the OSI and TCP/IP models run parallel to each other, often sharing the same functions and occasionally differing.

In this section we will describe and then compare the two models to illustrate the interactivity and similarity between them.

As we have seen, the TCP/IP model differs from the OSI model in that it consists of just four and not seven layers. The TCP/IP and the OSI models were developed concurrently, and each model has contributed to the development of the other.

The basic requirements for TCP/IP were defined as:

- A common set of applications
- Dynamic routing
- Connectionless protocols at the networking level
- Universal connectivity
- Packet switching

Figure 7.20 is a pictorial representation of this model shown in a similar fashion to the pictorial OSI model discussed earlier.

Note the placement of the TCP and IP components.

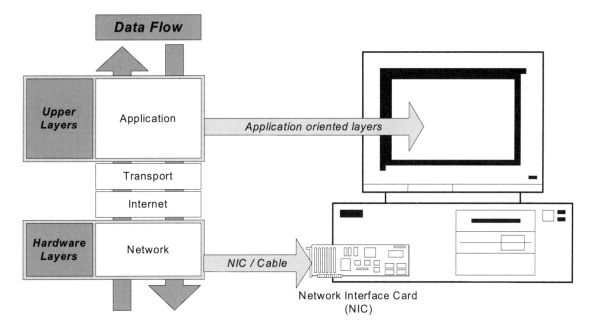

Figure 7.20
Pictorial view of the four TCP/IP layers

TCP/IP combines the (OSI) session and presentation layers into the TCP/IP application layer. In addition, because TCP/IP was required to provide a connectionless service, the (OSI) physical layer and data link layer were combined into the one TCP/IP network layer.

7.7.4 The Internet layers

The four layers, network, Internet, transport, and application (NITA) are described below. We start at the first layer, the network interface layer, which is sometimes referred to as the network access layer.

The network interface

This layer connects the local host to the local area network (LAN). Thus, it represents the physical and data link levels of the OSI model. It will use the required LAN operating algorithms, such as already defined above CMSA/CD or IBM token passing etc and is responsible for placing the data within a frame.

The frame format is dependent on the system being used, for example Ethernet LAN, frame relay, etc. The frame is the package that holds the data, in the same way as an envelope holds a letter.

The frame holds the hardware address of the host and checking algorithms for data integrity.

The ARPA model, because of its initial requirements under military use, offers good security to data. Some of the RFCs that apply to this layer are:

Asynchronous transfer mode (ATM)	Described in RFC 1438
Switched multimegabit data service (SMDS)	Described in RFC 1209
Ethernet	Described in RFC 894
IEEE 802 LANs	Described in RFC 1042
ARCNET	Described in RFC 1201
Serial line Internet protocol (SLIP)	Described in RFC 1055
Frame relay	Described in RFC 1490
Fiber distributed data interface (FDDI)	Described in RFC 1103
Packet switched public data networks (PSPDNs)	Described in RFC 877

The Internet layer

This performs the transfer of packets from one host to another. This is a 'packet' transfer, and not a frame transfer. A packet contains the address information needed for its routing through the internetwork to the receiving host.

The address within the frame header gets the frame from host-to-host on the same LAN. This layer is operated by the Internet protocol, the IP in TCP/IP.

However, there are several other additional protocols required at this level, these being:
- Address resolution protocol (ARP), RFC 826. The translation of an IP address to a network address, such as required by say Ethernet.
- Reverse address resolution protocol (RARP), RFC 903. This is the complement of ARP
- Internet control message protocol (ICMP), RFC 792. This is IP software used for communication from a gateway or host to its peers regarding any problems encountered. One of the best-known applications here is the Ping or Echo request that is used to test a communications link.

Host-to-host layer

This layer is primarily responsible for data integrity between the sender host and receiver host regardless of the path or distance used to convey the message. Communications errors are detected and corrected at this level. It has two protocols associated with it, these being:

User data protocol (UDP), which is a minimum security and basic protocol used for higher layer port addressing, defining the length and a checksum. It offers minimal protocol overhead. It is described in RFC 768.

Transmission control protocol (TCP), which has vastly improved protection and error control. This protocol, the TCP component of TCP/IP, is the heart of the TCP/IP suite of applications. It provides a very reliable method of transferring data in an 8-bit data format, known as octets, between two applications. This is described in RFC 793.

Process and application layer

The end-user interacts with the host via this layer. At this level there are many protocols used, some of the more common ones being:
- File transfer protocol (FTP), which as the name implies, is used for the transfer of files between two hosts using TCP. It is described in RFC 959.
- Trivial file transfer protocol (TFTP) is an economic version of FTP that uses UDP instead of TCP for reduced overhead. It is described in RFC 783.
- Simple mail transfer protocol (SMTP) is an example of e-mail application. It is described in RFC 821.
- TELNET (telecommunications network) is used to emulate terminals and for remote host access. It can, for example, emulate a VT100 terminal, across a network to a digital host.

Network management

The simple network management protocol (SNMP) provides a user with a common protocol, common to all platforms along with minimal overhead, that is used to communicate between the manager, or console, and the agent or device such as a bridge or router that is being managed. RFC 1157 details this application.

The leading network manufacturers such as IBM, Microsoft, Hewlett-Packard, SunSoft, and many others use this protocol.

7.8 PROFIBUS

7.8.1 Overview

PROFIBUS is a vendor-independent, open-field bus standard developed for the manufacturing and process automation environments. PROFIBUS uses communication profiles, namely DP and FMS. The physical profiles support RS-485, IEC 1158-2, and fiber optics. The PROFIBUS User Organization is currently developing vertical integration concepts for TCP/IP.

Application profiles define the technology required for individual device types and ensure vendor-independent device behavior.

7.8.2 Communication profiles

PROFIBUS communication profiles define the method in which users transmit data serially using the common transmission medium.

DP

DP is the most widely used communication profile in PROFIBUS and is optimized for speed, efficiency, and low connection costs. DP was designed as a replacement for conventional, parallel signal transmission with 24 volts in manufacturing automation as well as for analog 4–20 mA signals used in process automation.

FMS

FMS (fieldbus message specification) was designed for more sophisticated applications and demanding communication tasks between intelligent devices. FMS will become less significant in the future with the development of PROFIBUS towards TCP/IP integration.

7.8.3 Physical profiles

There are currently three transmission methods (physical profiles) available for PROFIBUS:
- RS-485 for applications in manufacturing automation
- IEC 1158-2 for use in process automation
- Optical fiber for eliminating interference and large distances

Further developments will make the use of PROFIBUS possible on commercial 10 Mbps and 100 Mbps Ethernet as physical layer.

RS-485

RS-485 is the transmission technology most frequently used by PROFIBUS. It is used in applications that demand a high transmission speed and straightforward, inexpensive installation. Twisted single pair shielded copper cable is used as physical medium.

One unique transmission speed has to be selected for all devices on the bus, selectable between 9.6 kb/s and 12 Mb/s.

All devices are connected in a physical bus structure. Up to 32 stations (masters or slaves) can be connected in one segment. Repeaters (line amplifiers) must be used to link up different bus segments if more than 32 stations are used. A total of 126 stations may be used in one network.

Type A cable is recommended for PROFIBUS using the RS-485 transmission technology, with the following parameters:

Impedance: 135 to 165 Ω
Capacity: < 30 pf/m
Loop resistance: 110 Ω/km
Wire gauge: 0.64 mm
Conductor area: > 0.34 mm^2

Using type A cable, the communication range per segment based on transmission speed will be as follows:

Baud Rate (kb/s)	9.6	19.2	93.75	187.5	500	850	1200
Range/Segment (m)	1200	1200	1200	1000	400	200	100

IEC 1158-2

IEC 1158-2 is used in process automation and specifically for hazardous areas. Synchronous transmission with a defined baud rate of 31.25 kbps is used, and the requirements for intrinsic safety and powering over the bus using two-wire technology are satisfied.

The use of PROFIBUS with IEC 1158-2 transmission technology in hazardous areas is prescribed by the FISCO model (fieldbus intrinsically safe concept). The FISCO model was developed by the Federal Physical Technical Institute in Germany, and is internationally recognized as the basic model for fieldbuses in hazardous areas.

Transmission in accordance with IEC 1158-2 and the FISCO model is based on the following principles:

- Each segment has only one source of power
- No power is fed to the bus while a station is sending
- Every field device consumes a constant basic current at steady state
- Field devices function as a passive current sink
- Passive line termination is performed at both ends of the main bus line
- Linear, tree and star topologies are allowed

Each station consumes a basic current of at least 10 mA in steady state. With bus powering, this current supplies energy to the field device. Communication signals are generated by the sending device by modulation of +/− 9 mA to the basic current.

A maximum number of 32 stations per line segment, with a total of 126 stations (4 repeaters) can be employed with IEC 1158-2. Line lengths of typically up to 1900 m are allowed.

7.8.4 Application profiles

PROFIBUS application profiles define the behavior of field devices during communication with PROFIBUS. The most important application profile is the PA profile, which defines the interaction of process automation devices. Application profiles are also available for variable speed drives, human–machine interfaces, etc.

7.8.5 Protocol architecture

PROFIBUS uses a multi-master system operating on the token bus access principle. The master devices (also called active stations) control the communication on the bus. A master can access the bus when it holds the token. Slave devices (also called passive stations) do not have bus access rights and can only acknowledge or send messages (only to the master) when requested to do so.

Figure 7.21 illustrates the PROFIBUS token ring concept.

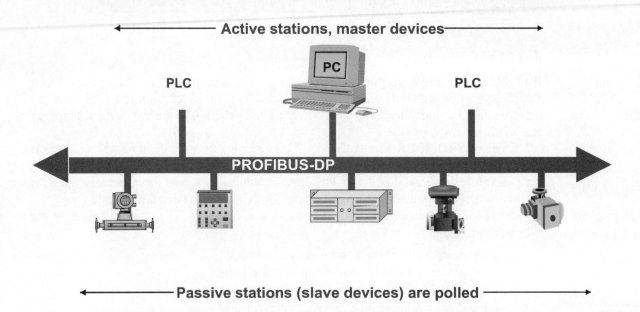

Figure 7.21
PROFIBUS token ring

The PROFIBUS protocol has been designed to meet two primary requirements for medium access control:

- During communication between two masters, it must be ensured that each of these stations gets sufficient time to perform its communication tasks within a precisely defined time interval, and
- For communication between a PLC and its slaves, cyclic, real-time data transmission needs to be implemented as fast and as simply as possible.

Therefore, the PROFIBUS medium access principle includes the token passing procedure, which is used by the masters to communicate with each other, and the master–slave procedure used by the master(s) to communicate with the slaves.

The token passing procedure ensures that the bus access right (token) is assigned to each master within a precisely defined timeframe. The token message must be passed around the logical token ring once to all masters within a maximum token rotation time (configurable). In PROFIBUS, the token passing procedure is only used for communication between masters.

The master–slave procedure permits the master that currently holds the token (active station) to access the assigned slaves.

This method of access allows implementation of the following system configurations:

- Pure master–slave system
- Pure master–master system (token passing)
- A combination of the above

A token ring is a logical ring of active stations via their bus addresses. The token, or bus access right, is passed from one master to the next in a predefined sequence of increasing addresses. When an active station receives the token, it can perform the master role for a certain period of time and communicate with all slave stations in a master–slave communication relationship and all master stations in a master–master relationship.

The medium access controller (MAC) of the active station will detect this logical assignment in the startup phase of the bus system and establish the token ring. The MAC is also responsible to detect defects on the transmission medium, errors in station addressing or in token passing.

The token hold time of a master will depend on the configured token rotation time and the number of active stations on the ring.

The protocol architecture is oriented to layers 1, 2 and 7 of the OSI model. The DP protocol uses layers 1 and 2, as well as a user interface, where the application functions and the device behavior of the various DP device types are specified. The direct data link mapper (DDLM) provides the user interface to layer 2.

The FMS protocol uses layers 1, 2 and 7. The lower layer interface (LLI) defines the representation of the FMS services on the data transmission protocol of layer 2. Layer 2 is called the field bus data link. Data security is also a task of layer 2.

The architecture of the PROFIBUS protocol referred to the OSI stack is shown in Figure 7.22.

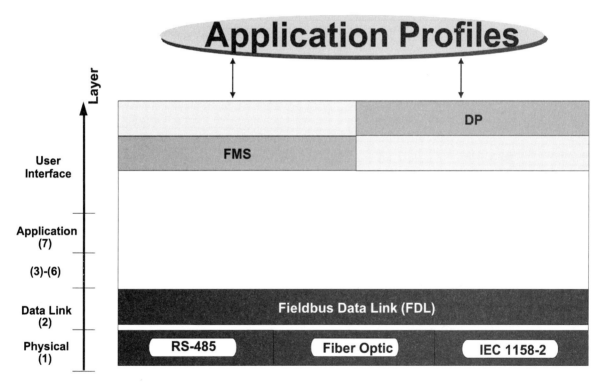

Figure 7.22
PROFIBUS protocol stack

PROFIBUS provides logical peer-to-peer data transmission, as well as multi-peer communication, namely broadcast and multicast.

In broadcast communication an active station sends an unacknowledged message to all other stations (masters and slaves).

Multicast communication means that an active station sends an unacknowledged message to a predetermined group of stations.

7.9 IEC 60870-5-101

7.9.1 Overview

The IEC 60870-5-101 protocol is defined with reference to layers 1, 2 and 7 of the ISO 7-layer stack, also called the enhanced performance architecture (EPA) model.

The protocol uses the octet representation of data. The frame is constructed as shown in Figure 7.23.

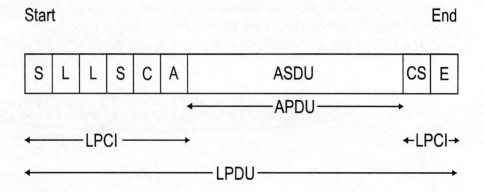

Figure 7.23
Frame construction

The ASDU is the data received from the application process. The ISO EPA model prescribes application protocol control information (APCI) to be added to the ASDU to form the application protocol data unit (APDU). However, this is not needed in the IEC 60870-5-101 protocol, hence the APDU is the same as the ASDU.

The data link layer adds the link protocol control information (LPCI) to the APDU to form the link protocol data unit (LPDU). This layer also prepares each data octet in the LPDU to be transmitted as an asynchronous serial character having one start bit (value=0), eight data bits (the data octet), one even parity bit and one stop bit (value=1).

Referring to Figure 7.23,
S = Start character
L = Length character
C = Link control character
A = Link address field
CS = Checksum character
E = End character

For up to 1.2 kb/s frequency shift keying (FSK) modulation is used, which is symmetrical and memory-less. It is suitable for most voice frequency analog channels on base band transmission line, power line carrier, and radio communications media.

For speeds in excess of 1.2 kb/s, synchronous modems may be used.

Faster transmission, up to 19.2 kb/s, is also possible on directly connected data circuits using digital signal multiplexers.

7.9.2 Frame integrity

The LPDU frame provides a very high data integrity (IEC Integrity Class I2). There must be at least four bit errors in a received frame before an undetectable frame error is possible.

The parity check ensures that at least two bit errors in the contents of any character are required to cause an undetectable character error.

The sum check ensures that at least two characters with undetected errors are required to cause an undetectable checksum error. Thus a total of four bit errors is required to produce a received frame with a possible undetectable error.

7.9.3 Physical configuration

The protocol may be implemented in two physical configurations:

a) A multidrop (bus) configuration, using master-slave communications with cyclic polling for media access. The access control will be performed on the data link layer.

b) Individual connections to all outstations in a typical star configuration. The individual connections permit balanced (full-duplex) media access for the link protocol, enabling spontaneous sending of data in both directions.

The protocol provides link functions for supporting full-duplex and half-duplex media access. However, practical considerations, including higher costs, may limit the extent to which full-duplex connections are used.

7.9.4 Application layer

The application layer of the protocol includes those functions that are concerned with communicating to application processes in a remote station. These functions are referred to as the 'user process' in the IEC 60870-5-101 standard.

a) Application functions

These include the following:

- Station initialization
- Data acquisition by polling
- Cyclic data transmission
- Acquisition of events
- Clock synchronization
- Acquisition of transmission delay
- Command transmission
- General interrogation
- Parameter loading
- Transmission of integrated totals
- Test procedure
- File transfer

Station initialization/interrogation

After the central control station and the various outstations have been initialized, the control station obtains an image of the present states of all digital and analog inputs. A station interrogation command is sent to all outstations, requesting that they return all their designated station interrogation data immediately.

The control station then constructs the network image, which is used as a starting point for future operations.

If, later, communication with an outstation is lost and then restored, a station interrogation command may be used to obtain a static update of only the part of the network image belonging to that specific outstation.

Acquisition of events

Once the network image has been constructed, the image must be updated with dynamic data obtained from the acquisition of events function, if and when changes take place in the network.

This dynamic data is usually vital input information, which may require immediate action at the control station.

For example, a critical circuit-breaker may change state. The outstation application layer presents a request ASDU to its link layer, which ensures that the ASDU is transmitted to the control station. The control station link layer will, upon receipt of the data, generate an indication to its application layer, presenting the received ASDU. The ASDU will be marked with 'cause of transmission = spontaneous', which prompts the application layer to immediately update the specific point in the network image. Subsequently, other urgent operations may take place as a result, for example activating an alarm.

Optional double transmission of spontaneous events may be configured if required. The first transmission takes place as described above. A second lower priority transmission of the same event, with an added time tag, may be used for an event record, to be analyzed at a later stage to determine the exact sequence in which events occurred.

Background scan

This function is a slow cyclic scan, which is used to ensure that the network image values are up-to-date, and that no spontaneously reported events went undetected since the last scan cycle.

Cyclic transmission

This function is used to continuously update another central image with information data obtained from measurements taken at regular intervals of time. It is used for monitoring the less critical inputs, i.e., either slow changing values or changes that do not require fast action at the central station, for example a transformer temperature.

The ASDU received by the central station will be marked with 'cause of transmission = cyclic', which will prompt relatively slow actions including the updating of the related points in the central image and cyclic updating of an operator display.

Transmission of integrated totals

Integrated totals, for example kilowatt-hours, are accumulated by outstations. The accumulated total in a register may be sampled using a freeze command, after which the total may continue to accumulate or start again from zero by using a reset command.

This may be achieved by one of the following modes of operation:

Mode A: The freeze and reset commands are initiated from time signals from the outstation local clock. The resulting integrated totals or incremental information are transmitted to the control station in ASDUs with 'cause of transmission = spontaneous'.

Mode B: The freeze and reset commands are initiated from time signals from the outstation local clock. The resulting integrated totals or incremental information are held at the outstation until they are requested by an interrogation command from the control station. The responding ASDUs have 'cause of transmission = requested'.

Mode C: Interrogation commands are sent periodically to outstations requesting integrated totals or incremental information. The freeze and reset commands are incorporated within these interrogation commands from the control station. The responding ASDUs have 'cause of transmission = requested'.

Mode D: The freeze and reset commands are sent periodically to outstations. The integrated totals or incremental information are returned spontaneously.

Command transmission

ASDUs containing commands are sent from the control station to the outstations when required. These will include commands for changing the state of digital (on/off) outputs, step (raise/lower) outputs and set point (analog) outputs, as well as interrogation commands and clock synchronization commands. Two modes of operation are provided: Direct (immediate) execution and select/execute, where the selection is confirmed back before the actual execution ASDU is sent.

File transfer

Files are generally held in the part of the system where they are generated, until they are transferred to a central data storage location. For example, disturbance records may be held in the protection equipment, and event records in the control equipment, of an electrical power substation. To enable efficient file transfer, use is made of directories and subdirectories.

The control station may request a directory/subdirectory from an outstation, to inform it what files are currently available for transfer from the outstation, or the contents of a directory/subdirectory may be transmitted spontaneously to the control station when changes occur.

Disturbance records contain data in a format that is easily mapped onto the format of the IEC 60870-5-103 protocol (informative interface for protection equipment).

Event records from an outstation may be used instead of, or additional to, the 'acquisition of events' application, as described earlier.

b) Application service data units

Clock synchronization

The protocol provides the facility to synchronize the clocks in all outstations to the clock in the control station, to ensure accuracy of time tags. Allowance for the transmission time delay of the command message may be made.

The clock synchronization function should not be used if constancy of transmission time delay cannot be guaranteed.

The protocol offers different types of ASDU suitable for the specific application. However, they all have the same general format as shown in the Figure 7.24.

Figure 7.24

T = Type identification (1 data octet)
Q = Variable structure qualifier (1 data octet). This indicates the number of information objects in the ASDU
C = Cause of transmission (1 or 2 data objects). Causes may be: cyclic, spontaneous, request/requested, activate (control action), etc
CA = Common address (1 or 2 data objects)
OA = Information object address (1, 2 or 3 data octets)
IE = Set of information elements (as defined for the type of ASDU specified in the T field)
TT = Time tag of information object

7.9.5 Summary

The IEC protocol is mainly concerned with standardizing specifications so that different suppliers of stations can conform to a common set of provisions for a particular installation, to ensure interoperability.

No peer-to-peer communications are defined in the standard, and no detailed polling process. The standard only defines the sequence in which the standard link functions are used to acquire the application data waiting for transfer to the central station.

7.10 IEC 60870-5-103

7.10.1 Overview

The IEC 60870-5-103 has been termed a 'companion standard for the informative interface of protection equipment'. DNP V3.0 was based on an early version of this standard.

The physical layer supports two physical configurations: RS-485 for the electrical interface and fiber optics for the optical interface. Transmission speeds of either 9.6 kb/s or 19.2 kb/s are available.

The protocol operates as a master–slave system and uses cyclic polling as the medium access method. Layers 1, 2 and 7 of the OSI model are defined.

7.10.2 Standard information numbers

Standard information numbers are defined for information, monitor, and control purposes, as follows:
(Extracted from 60870-5-103 © IEC: 1997, page 163–169)
 Selection of standard information numbers in monitor direction

System functions in monitor direction

	INF	Semantics
☒	<0>	End of general interrogation
☒	<0>	Time synchronization
☒	<2>	Reset FCB
☒	<3>	Reset CU
☒	<4>	Start/restart
☒	<5>	Power on

Status indications in monitor direction

	INF	Semantics
☒	<16>	Auto-recloser active
☒	<17>	Teleprotection active
☒	<18>	Protection active
☒	<19>	LED reset
☒	<20>	Monitor direction blocked
☒	<21>	Test mode
☒	<22>	Local parameter setting
☒	<23>	Characteristic 1
☒	<24>	Characteristic 2
☒	<25>	Characteristic 3
☒	<26>	Characteristic 4
☒	<27>	Auxiliary input 1
☒	<28>	Auxiliary input 2
☒	<29>	Auxiliary input 3
☒	<30>	Auxiliary input 4

Supervision indications in monitor direction

	INF	Semantics
☒	<32>	Measurand supervision I
☒	<33>	Measurand supervision V
☒	<35>	Phase sequence supervision
☒	<36>	Trip circuit supervision
☒	<37>	I>> backup operation
☒	<38>	VT fuse failure
☒	<39>	Teleprotection disturbed
☒	<46>	Group warning
☒	<47>	Group alarm

Earth fault indications in monitor direction

INF	Semantics
☒ <48>	Earth fault L_1
☒ <49>	Earth fault L_2
☒ <50>	Earth fault L_3
☒ <51>	Earth fault forward, i.e. line
☒ <52>	Earth fault reverse, i.e. busbar

Fault indications in monitor direction

INF	Semantics
☒ <64>	Start/pick-up L_1
☒ <65>	Start/pick-up L_2
☒ <66>	Start/pick-up L_3
☒ <67>	Start/pick-up N
☒ <68>	General trip
☒ <69>	Trip L_1
☒ <70>	Trip L_2
☒ <71>	Trip L_3
☒ <72>	Trip I>> (backup operation)
☒ <73>	Fault location X in ohms
☒ <74>	Fault forward/line
☒ <75>	Fault reverse/busbar
☒ <76>	Teleprotection signal transmitted
☒ <77>	Teleprotection signal received
☒ <78>	Zone 1
☒ <79>	Zone 2
☒ <80>	Zone 3
☒ <81>	Zone 4
☒ <82>	Zone 5
☒ <83>	Zone 6
☒ <84>	General start/pick-up
☒ <85>	Breaker failure
☒ <86>	Trip measuring system L_1
☒ <87>	Trip measuring system L_2
☒ <88>	Trip measuring system L_3
☒ <89>	Trip measuring system E
☒ <90>	Trip I>
☒ <91>	Trip I>>
☒ <92>	Trip IN>
☒ <93>	Trip IN>>

Auto-reclosure indications in monitor direction

INF	Semantics
☒ <128>	CB 'on' by AR
☒ <129>	CB 'on' by long-time AR
☒ <130>	AR blocked

Measurands in monitor direction

INF	Semantics
☒ <144>	Measurand I
☒ <145>	Measurands I, V
☒ <146>	Measurands I, V, P, Q
☒ <147>	Measurands I_N, V_{EN}
☒ <148>	Measurands $I_{L1,2,3}$, $V_{L1,2,3}$, P, Q, f

Generic functions in monitor direction

INF	Semantics
☒ <240>	Read headings of all defined groups
☒ <241>	Read values or attributes of all entries of one group
☒ <243>	Read directory of a single entry
☒ <244>	Read value or attribute of a single entry
☒ <245>	End of general interrogation of generic data
☒ <249>	Write entry with confirmation
☒ <250>	Write entry with execution
☒ <251>	Write entry aborted

Selection of standard information numbers in control direction

System functions in control direction

INF	Semantics
☒ <0>	Initiation of general interrogation
☒ <0>	Time synchronization

General commands in control direction

INF	Semantics
☒ <16>	Auto-recloser on/off
☒ <17>	Teleprotection on/off
☒ <18>	Protection on/off
☒ <19>	LED reset
☒ <23>	Activate characteristic 1
☒ <24>	Activate characteristic 2
☒ <25>	Activate characteristic 3
☒ <26>	Activate characteristic 4

Generic functions in control direction

INF	Semantics
☒ <240>	Read headings of all defined groups
☒ <241>	Read values or attributes of all entries in one group
☒ <243>	Read directory of a single entry
☒ <244>	Read value or attribute of a single entry

☒ <245> General interrogation of generic data
☒ <248> Write entry
☒ <249> Write entry with confirmation
☒ <250> Write entry with execution
☒ <251> Write entry abort

Basic application functions

☒ Test mode
☒ Blocking of monitor direction
☒ Disturbance data
☒ Generic services
☒ Private data

7.11 UCA 2.0 (an overview)

The Electric Power Research Institute (EPRI) developed the suite of protocols known as Utilities Communication Architecture, Version 2 (UCA 2.0). This protocol is based on the Ethernet physical and link layers, incorporates the TCP/IP collection of protocols, and utilizes the MMS protocol for the application layers.

The protocol defines all seven layers of the OSI model. The Ethernet type CSMA/CD media access principle is employed, and full peer-to-peer communications, as well as master–slave communications, are supported.

The protocol is intended to be a globally recognized standard to provide an open, integrated method for exchanging real-time data between utility systems. It may be utilized within the substation environment and on an enterprise level.

UCA offers interconnectivity and interoperability between equipment from different manufacturers to exchange high-speed, real-time data in utility operations.

UCA is based on various international protocols and standards. The first version, developed in 1991, provided the necessary communication requirements and guidelines for use. UCA Version 2, completed in 1998, provides additional communication profiles, application services, and device models for interoperability among various equipment.

The EPRI submitted UCA 2.0 to the Institute of Electrical and Electronic Engineers (IEEE) for publication as a Technical Standard under the IEEE Standards Board.

More information regarding UCA can be obtained on the EPRI website www.epri.com.

7.12 Standardization

The OSI model provided an invaluable reference standard for protocols, with the result that protocol converters became viable. This was the immediate solution to enable equipment from different manufacturers using different protocols to communicate on the same network. However, it was realized that a need existed for standardization of communication protocols for specific applications.

The IEC Technical Committee (TC) 57 was given the task to prepare international standards for power system control equipment and systems. Various working groups within TC57 are setting standards for different levels of communications.

Two important standards have been accepted up to date, namely IEC 60870-5-103, which is a 'Companion Standard for the Informative Interface of Protection Equipment', and IEC 60870-5-101, which is a standard for data communications on a network level.

However, it is realized that IEC 60870-5-103 is not supportive of interoperability between IEDs in a substation. The intended standard for power system automation systems, to define the communications between IEDs and the related system requirements, is IEC 61850. The specific goal of IEC 61850 is interoperability, which has been defined as the ability of two or more IEDs from the same vendor or different vendors to exchange information and use that information.

Simultaneously, the Electric Power Research Institute (EPRI), was developing a similar standard for the USA market and the protocol UCA 2.0 has been introduced (Utility Communication Architecture), and accepted as a standard by the IEEE. The IEC and IEEE agreed in February 1998 to work towards one international standard, hence UCA 2.0 will probably be integrated into the IEC 61850 standard.

8

SCADA systems

8.1 Definition and background

The term SCADA is the abbreviation for supervisory control and data acquisition. It encompasses the collection of information (data acquisition), transferring the data over physical mediums (the field of telemetry/data communication) and the processing and display of the data at the master station. The master station is also used for centralized control over the communication network and/or to initiate external commands (supervisory control).

The term SCADA master is commonly used to refer to the master station, including the hardware and software. This will be the emphasis of this chapter, as the concepts of data acquisition and communication, as applicable to power system automation, has already been discussed comprehensively in other chapters.

SCADA masters were originally only intended to communicate to remote terminal units (RTUs), as the master station and the RTU formed the core of a comprehensive SCADA philosophy. Thereafter, SCADA software was developed to communicate to programmable logic controllers (PLCs) and nowadays SCADA software is available that can communicate to virtually any device capable of advanced communications.

The early SCADA masters were solely manufacturer-based software, as well as hardware, with no interchangeability between products from different manufacturers. Fortunately, the tremendous development and popularity of the personal computer (PC) actually forced suppliers of SCADA software to develop their products to be compatible with PC use. Nowadays all leading SCADA software suppliers' products will run on a PC.

SCADA software applicable to power system automation differs in their design base. Some SA suppliers use their own proprietary software, specifically designed for their systems, whereas other suppliers use 'open' SCADA software that was developed by independent SCADA software developers, and which are generally able to support a multitude of communication protocols.

8.2 Requirements for the SCADA master station

The SCADA master station should perform the following functions in a power system automation system:

- Display real-time data received from the IEDs, relays, bay controllers, RTUs or PLCs connected in the power system
- Keep historical records of data received and retrieve these records when required
- Activate alarms when necessary
- Display sequence-of-event reports and disturbance recordings when required
- Provide an active operator interface for supervisory control and remote configuration of IEDs and other devices
- Perform communication control over the network, depending on the protocol used

8.2.1 Hardware

A SCADA master station may be utilized to control one substation only, as will be the case in large generation or transmission high-voltage substations. The station will then probably form part of a LAN or WAN network. This configuration is illustrated in Figure 8.1.

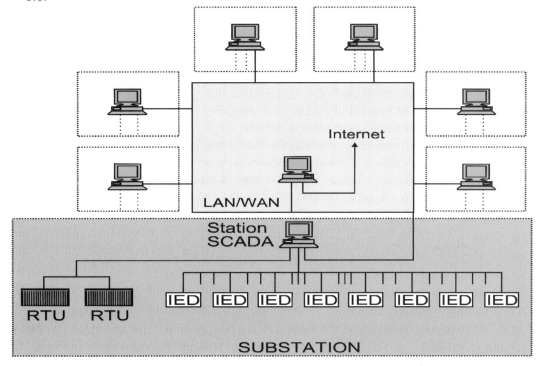

Figure 8.1
Network with substation SCADA station

Smaller distribution substations will normally not justify a SCADA station for each substation, and several substations will be connected to one SCADA master station, as illustrated in Figure 8.2.

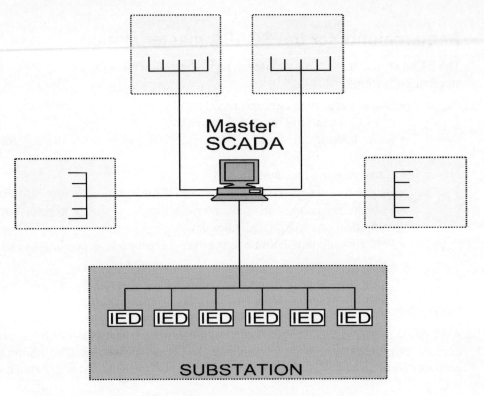

Figure 8.2
Master SCADA station for several substations

The hardware for a power system automation master station will normally consist of one or more desktop PCs or workstations, with its normal peripheral devices and one or more communication ports. The reliability of the hardware is crucial. Electrical protection functions will not, or should not, be dependent on the availability of the SCADA master or the communications network. Likewise, most SA IEDs are able to store large amounts of data, which can be retrieved again should the link be temporarily lost. However, real time data acquisition and remote control will be unavailable in case of hardware failure. This may be catastrophic in a SA system, depending on the criticality of the power system.

Therefore, a high-end and reliable PC (or workstation) should be installed for the SCADA master. The cost of the PC hardware for the SCADA master is a relative low component of the total SA system cost, but it is the single most crucial element of the whole system. It will not make sense at all to attempt to save money on this part of the system. Install the best and most reliable PC available on the market – it will be well worth the investment.

It is good practice to utilize two PCs at the master station. The first PC then functions as the 'operator terminal'. This PC will normally be the SCADA master, and all remote control commands will be initiated by the operator from this terminal. The second PC functions as the 'engineering terminal'. This PC has two main functions – firstly for redundancy, to take over as SCADA master should the first PC fail; and secondly to retrieve and view disturbance records, sequence-of-event recordings, change relay settings and/or configurations, etc, without interfering with the operator functions. This could be especially useful in abnormal power system conditions, where many things may be happening at the same time.

A great deal of attention should be given to the power supply to the PC(s). A failure of the main power supply usually means abnormal power system conditions somewhere, and this is precisely when the power system automation system is needed the most. The most reliable way is to provide a dedicated UPS (uninterruptable power supply) to the PC(s). The UPS should be able to provide power for several hours if necessary, until power system conditions return to normal. It is always good practice to feed the UPS with two power supplies – the normal supply as well as a backup or emergency supply if available. The UPS should be monitored constantly by the SCADA system and an early warning given should the UPS become unhealthy.

The normal peripheral devices will be required, like a keyboard, mouse, etc, as well as a data comms port. A CD writer or digital tape drive should be installed to regularly back-up and archive data. A dot-matrix printer could be installed at the operator terminal to print events as they occur, and a laser or ink-jet printer at the engineering terminal to print reports. (This will depend on the operating philosophy of the client.)

8.2.2 Software

Three software platforms are applicable to the SCADA master station for a power system automation system, namely:

- The operating system software
- The SCADA system software
- The SCADA application software

8.2.2.1 Operating system software

Reliable, stable software should be chosen for the operating system. It should be ascertained on which platform the SCADA system software can run and perform the best.

A detailed discussion of the operating software falls outside the scope of this text.

8.2.2.2 SCADA system software

The SCADA system software will usually be supplied as a standard package and then configured for or by the particular user. Some suppliers' power system automation systems will utilize 'open' SCADA software, which is supplied by independent SCADA software developers, like Wonderware and Citect. This software will work well with power system automation equipment, but it must be kept in mind that they were developed primarily for process automation applications, and may be limited when it comes to power system automation applications.

There are SCADA software suppliers that develop independent, 'open' software specifically for power system automation, for example Telegyr in Sweden. These software tend to be more expensive, as their market is more limited, but it may be well worth the investment, as they have been developed, and tested thoroughly, exclusively with power system automation applications in mind.

Other suppliers of power system automation systems use their own developed SCADA software; for example ABB's MicroSCADA, Altom's MiCOM system, etc. This software has been developed for that specific manufacturer's own products, and is generally more expensive than the 'common' SCADA software. It is not to say that these manufacturers' equipment may not be used with an 'open' SCADA package, for example using ABB relays with Citect SCADA software. However, often protocol conversions become necessary, and some functionality may be lost upon the way, as it is maybe not supported by the specific SCADA software.

There are typically five tasks in any SCADA system. Each of these tasks performs its own separate processing:

- Input/output task
  This program is the interface between the control and monitoring system and the plant floor.
- Alarm task
  This manages all alarms by detecting digital alarm points and comparing the values of analog alarm points to alarm thresholds.
- Trends task
  The trends task collects data to be monitored over time.
- Reports task
  Reports are produced from plant data. These reports can be periodic, event triggered or activated by the operator.
- Display task
  This manages all data to be monitored by the operator and all control actions requested by the operator.

Generally, the SCADA system software consists of four main modules:
- Data acquisition
- Control
- Archiving/database storage
- The human–machine interface (HMI)

Data acquisition

Data acquisition refers to the reception, analyzing and processing of all data from the field. The real-time data will normally be graphically displayed, according to the user-defined configuration. Data will be compared to pre-set limits, and alarms raised when these limits are exceeded. Some alarms will be identified and flagged by the field device and others by the SCADA master, depending on the specific system and the configuration.

The way data is accessed from the field devices will depend on the system configuration and communication protocol. This will determine if the SCADA master software actively and continuously control the communication network, or if it only acts as an information and remote control center.

Control

Control commands from the SCADA master station to the field devices will either be initiated by the operator, or automatically according to pre-defined parameters.

Archiving/database storage

The way in which historical data is stored and archived will depend on the hardware, software, as well as user configuration. Historical data need to be assessed for trend information, fault finding, and reporting. The data backup and storage philosophy needs careful consideration by the user.

The human–machine interface

The human–machine interface (HMI) consists of the input and output devices for interaction between the operator and the software, and the way the data is displayed or commands given to the system. Output will normally be on a graphic display screen for real-time data, and to a printer for event records and reports. Input will generally be via a keyboard and mouse. Although convenient, touch screen technology should be viewed with caution for power system automation, due to security considerations. It is too easy to accidentally give a wrong command with a touch screen, which can be disastrous.

A well-designed SCADA display should have different levels to graphically display the real-time situation of the plant. The first level should present an overview of the plant to be monitored, as illustrated in Figure 8.3. The substations that are monitored are shown with some important values to give an overall view of the plant. The graphic display screens are totally user-configurable. Different colors will be used to differentiate different voltages, healthy or alarm conditions, etc. An alarm will change the color of the substation where it is coming from, maybe blinking and/or sounding an audible alarm, depending on user preference. The display can automatically switch to the alarmed device, or leave it to the operator to do so.

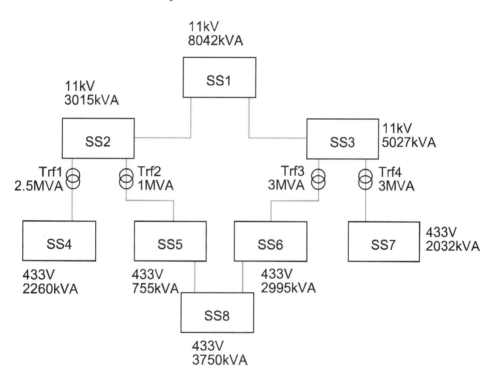

Figure 8.3
Example of SCADA overview screen

The second level should present the detail of the relevant substation, as illustrated in Figure 8.4. The substation in the example is a double busbar system, but only superficial detail is shown with the most important electrical values that should be visible with one glance. Different colors will be used to indicate breaker open/close status and alarm conditions.

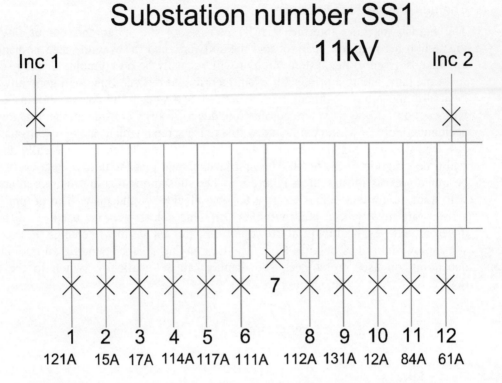

Figure 8.4
Example of second level screen

A third level should present the required detail of the selected feeder, as illustrated in Figure 8.5. Every detail and all electrical values required should be presented on this level. Detail of alarm conditions will also be displayed here. The reason why all detail should be presented on this level is simply not to clog the higher level displays with too much data. Control commands, for example open or close a circuit-breaker, will ideally be done from this level. Bay level interlocking information will be displayed here, while station level interlocking information will be displayed on the next higher level.

A fourth level may be utilized, for example to open a single device information window, eg. a circuit-breaker, where condition-monitoring information may be displayed.

The use of passwords should be carefully considered for security purposes. Most SCADA software supports the application of multiple passwords, to control access to different levels and commands.

Graphics should be kept as clear as possible, displaying all important information while avoiding cluttering the display with too much information. Rather use more levels than providing too much detail on one display. Access to different levels is normally obtained by typical Windows-style commands, clicking the mouse pointer on the desired object to display.

Operator commands should include at least one confirmation step, preferably more for critical commands, to avoid accidental and very costly mistakes.

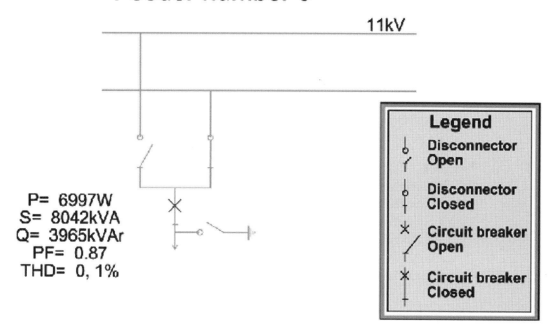

Figure 8.5
Example of detail screen

8.2.2.3 SCADA application software

SCADA application software is typically optional packages to add on enhanced features to the main SCADA software. Examples of this would be software tools for analyzing disturbance recordings, change protection settings and configuration of a specific manufacturer's relays remotely, report writing software, etc.

The SCADA application software need not be from the same supplier as the SCADA main software, but needs to be compatible and at least run on the same operating system software. Different levels of integration will be applied between the SCADA operating software and application software, depending on the specific software and the tasks it is required to perform.

8.2.3 General SCADA features

The following are some general features that should be present in any well-designed SCADA system, and which is also applicable to power system automation:

8.2.3.1 Alarms

- Client server architecture
- Time stamped alarms to 1 millisecond precision (or better)
- Single network acknowledgment and control of alarms
- Alarms are shared to all clients
- Alarms displayed in chronological order

- Dynamic allocation of alarm pages
- User-defined formats and colors
- Two to four adjustable trip points for each analog alarm
- Deviation and rate of change monitoring for analog alarms
- Selective display of alarms by category (256 categories)
- Historical alarm and event logging
- Context-sensitive help
- On-line alarm disable and threshold modification
- Event-triggered alarms
- Alarm-triggered reports
- Operator comments can be attached to alarms

8.2.3.2 Trends

- Client server architecture
- True trend printouts
- Rubber band trend zooming
- Export data to DBF, CSV files
- X/Y plot capability
- Event based trends
- Pop-up trend display
- Trend gridlines or profiles
- Background trend graphics
- Real-time multi-pen trending
- Short and long term trend display
- Length of data storage and frequency of monitoring can be specified on a per-point basis
- Archiving of historical trend data
- On-line change of time-base without loss of data
- On-line retrieval of archived historical trend data
- Exact value and time can be displayed
- Trend data can be graphically represented in real-time

8.2.3.3 IED interface

- All required protocols included
- DDE drivers supported
- Interface also possible for RTUs, loop controllers, bar code readers and other equipment
- Driver toolkit available
- Operates on a demand basis instead of the conventional predefined scan method
- Optimization of block data requests to IEDs
- Rationalization of network user data requests
- Maximization of comms bandwidth
- Additional hardware can be added without replacing or modifying existing equipment
- Limited only by the IED architecture

8.2.3.4 Access to data

- Direct, real-time access to data by any network user
- Third-party access to real-time data, e.g. Lotus 123 and EXCEL
- Network DDE
- DDE compatibility: read, write and exec
- DDE to all IO device points
- Clipboard
- ODBC driver support
- Direct SQL commands or high level reporting

8.2.3.5 Networking

- Supports all NetBIOS compatible networks such as NetWare, LAN Manager, Windows for Workgroups, Windows NT
- Support protocols NetBEUI, IPX/SPX, TCP/IP and more
- Centralized alarm, trend and report processing – data available from anywhere in the network
- Dual networks for full LAN redundancy
- No network configuration required (transparent)
- May be enabled via single check box, no configuration
- LAN licensing is based on the number of users logged onto the network, not the number of nodes on the network
- No file server required
- Multi-user system, full communication between operators
- RAS and WAN supported with high performance
- PSTN dial-up support

8.2.3.6 Client/server distributed processing

- Open architecture design
- Real-time multitasking
- Client/server fully supported with no user configuration
- Distributed project updates (changes reflected across network)
- Concurrent support of multiple display nodes
- Access any tag from any node
- Access any data (trend, alarm, report) from any node

8.2.4 Other considerations

8.2.4.1 System response times

These should be carefully specified for the following events. Typical speeds that are considered acceptable are:

- Display of analog or digital value (acquired from IED) on the master station operator display (1 to 2 seconds maximum)
- Control request from operator to IED (1 second critical; 3 seconds non-critical)
- Acknowledge of alarm on operator screen (1 second)

- Display of entire new display on operator screen (1 second)
- Retrieval of historical trend and display on operator screen (2 seconds)
- Sequence of events logging (at IED) of critical events (1 millisecond)

It is important that the response is consistent over all activities of the SCADA system. Hence the above figures are irrelevant unless the typical loading of the system is also specified under which the above response rates will be maintained. In addition *no loss of data must occur* during these peak times.

A typical example of specification of loading on a system would be:

- 40% of all digital points change status every 10 seconds (or go from healthy into alarm condition).
- 60% of all analog values undergo a transition from 40 to 100% every 5 seconds.

8.2.4.2 Expandability of the system

Whilst performance and efficiency of the SCADA package with the current plant is important, the package should be easily upgradable to handle future requirement. The system must be easily modifiable as the requirements change, and expandable as the task grows – in other words the system must use a scaleable architecture.

A typical figure quoted in industry is that if expansion of the SCADA system is anticipated over the life of the system the current requirements of the SCADA system should not require more than 60% of the processing power of the master station and that the available mass storage (on disk) and memory (RAM) should also be approximately 50% of the required size.

It is important in specifying the expansion requirements of the system that:

- The additional hardware that will be added will be of the same modular form as that existing and will not impact on the existing hardware installed.
- The existing installation of SCADA hardware/control cabinets/operator displays will not be unfavorably impacted on by the addition of additional hardware. This includes items such as power supply/air conditioning/SCADA display organization.
- The operating system will be able to support the additional requirements without any major modifications.
- The application software should require no modifications in adding new IEDs or operator stations at the central site/master station.

9

Communications in power system automation

9.1 Overview

Power system automation, as defined in this manual, is illustrated in Figure 9.1. Data communications is the core of a *power system automation* system. Without communications the whole system will collapse. The role and requirements of communications for each segment in the context of power system automation is now analyzed.

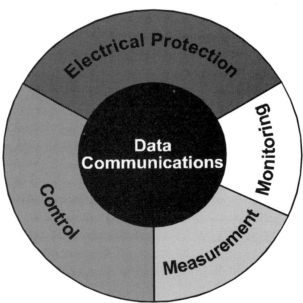

Figure 9.1
Power system automation

9.2 Configuration

The typical power system automation configuration is illustrated in Figure 9.2.

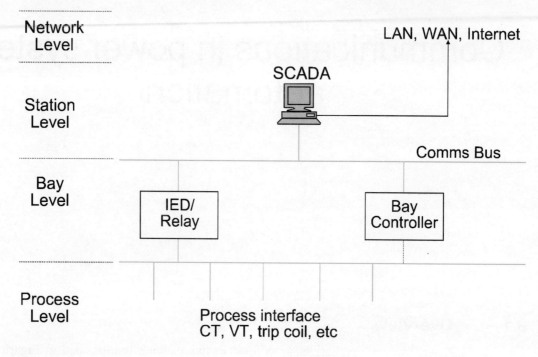

Figure 9.2
Typical power system automation structure

- The process level consists of the equipment providing information to the bay level, e.g. instrument transformers, temperature sensors, auxiliary contacts of circuit-breakers, etc. The application process is therefore a voltage, current, temperature, breaker status, etc.
- The equipment executing a command from the bay level, e.g. trip coil of circuit-breaker. The application process is then the command 'open breaker'.
- The bay level consists of four main application processes (APs): protection, control, measurement/metering, and monitoring. These APs can reside in different devices, or all in one device (the typical IED), as discussed in Chapter 5.
- The station level consists of the station SCADA (optional) and possibly a gateway or communications processor. The importance of the station SCADA will depend on the specific application. In large transmission substations, this will form the main SCADA for the specific substation, with several SCADA systems forming a network. On the other hand, for a distribution substation, the station SCADA may be dispensed with, and only a gateway will be required to connect the substation to the network and to the main SCADA.

- The gateway is usually some form of communications processor, star coupler or such, depending on the vendor's system.
- The network level may consist of a central SCADA, to which each substation is connected, and/or a LAN, MAN, WAN or the Internet.

9.3 Communication requirements

The communication requirements for the various applications in power system automation will be evaluated in this section according to the following attributes:

Performance	Low	Medium	High
Speed/Data throughput	<10 kbps	>10 kbps < 1 Mbps	>1 Mbps
Response time	>1 s	<1 s >10 ms	< 0 ms
Time Synchronization	1 s	1 ms ± 0.1 ms	1 µs ± 0.5 µs
Avalanche Handling	No data through-put required during avalanche	Some data through-put required	All data through-put required
Data Integrity	Some errors allowed	Limited errors allowed	No errors allowed
Link Availability	All data can wait until link available	Some data can wait for a limited time	No data can wait
Data Priority	Can be sent after all other data	To be sent after high priority data; can wait for request	To be sent immediately; cannot wait for request

Table 9.1
Communication requirements

9.3.1 Protection

Electrical protection has always functioned independently on the bay level, and that is how it should stay. Protection needs to be fast, reliable and secure. In the early 1990s there have been some proposals to move protection to a network level. Thankfully, it was realized, mainly by protection engineers, that a network level could not support the stringent requirements of protection, and these proposals died a natural death.

Nowadays, with communication networks becoming faster, more reliable, more secure and more powerful by the day, there is again periodic, although tentative, suggestions to consider providing protection on a network level. Fortunately, no switchgear or relay manufacturer is seriously considering this at present.

Protection functions have been traditionally provided by electromechanical relays, with one relay dedicated for one function. Digital relays opened the way for multi-function relays, with more than one protection function in one device. Today, IEDs provide protection functions together with control functions, measurement, monitoring and communications. This is making some protection engineers nervous, and many of them prefer to have protection and control functions in different devices. The advantages and disadvantages of this approach are discussed in detail in Chapter 11.

The fact is that protection is often not the dedicated function of one or more devices anymore; therefore it is more appropriate to view protection as an application, rather than a hardware device.

A. Lower Level

The field information to the protection application is provided by the instrument transformers (mainly) and other electrical transducers. This information has always been provided as analog values in the form of current and voltage. The instrument transformers and other equipment are hardwired to the input terminals of the protection device.

The concept of 'smart instruments', i.e. field instruments that have local intelligence and advanced communication capabilities, has also influenced the electrical industry. Instrument transformers and other electrical transducers have been developed that convert the analog values into a data message, to be sent over a serial communication bus to the protection AP (typically using fiber-optic).

The following are the advantages and disadvantages of having a digital communication bus between the process level and the bay level, compared to the conventional hardwired method:

Advantages

- Huge reduction in wiring, as all the input information can be sent via one fiber-optic cable, replacing a multitude of copper wiring.
- Eliminating the risk of faulty wiring and/or loose connections.
- Reducing the risk of human error, e.g. to open-circuit a CT.
- The field information can be easily made available to more than one AP (instead of having, for example, one set of CTs for each function).
- Eliminating the risk that high electrical transients can damage equipment.
- Intelligent field information bus with self-supervision and fault diagnostic capabilities.

Disadvantages

- More complex to configure and maintain
- Higher risk that 'something can go wrong', compromising the reliability of the protection scheme
- Inherent time delay
- Compatible equipment required
- Higher initial cost

Although data communications do not commonly apply to the process or lower level as yet regarding electrical substations (as this level constitutes mainly the direct electrical values of voltages and currents), the very near future will see data communications being utilized on this level more and more.

Therefore, it is appropriate to examine the requirements that will apply to this level. Herewith follows the requirements for communications on the lower level concerning the protection application:

Requirements

Performance	Level required
Speed/Data throughput	High
Response time	High
Time synchronization	High
Avalanche handling	High
Data integrity	High
Link availability	High
Data priority	High

Notes:
- Very high response times are required for some protection schemes, e.g. differential and distance protection.
- Very high time synchronization is required for distance protection, synchronizing check, etc.

As can be seen, a high level (minimum) is required from all the performance attributes. This is due to the fact that immediate and absolute reliable information is needed for the protection scheme to function correctly, effectively and reliably.

This stringent requirements for this communication level will probably see to it that conventional wired systems will still be predominantly used for a long time, especially in distribution applications, both due to high initial cost and scepticism among protection engineers that the requirements will not be met by having a data communication link on this level.

Protection schemes like differential protection, busbar protection, etc must rely on point-to-point communications and not communications in a 'public domain'.

B. Higher level

This is the communication level from the protection AP to the station or central SCADA. The protection AP performs its functions independent from the SCADA. Information send to the SCADA will include disturbance records, event records, and status information. Information from the SCADA to the protection AP will include program configuration and protection settings data.

Requirements

Performance	Level required
Speed/Data throughput	Low
Response time	Low
Time synchronization	High
Avalanche handling	Low
Data integrity	High
Link availability	Low
Data priority	Low

Explanatory notes:
- The requirements for speed, response time, avalanche handling, data priority and link availability are low, as the information required is stored in the IED and can be retrieved again. The information is usually not of an urgent nature. Similarly, data from the SCADA to the IED, like protection settings, is not high priority.

- Time synchronization is high to aid in sequence-of-event recordings.
- Data integrity is high as no corruption of critical data, like protection settings, can be allowed.

9.3.2 Control

Traditionally, substation control has been utilized exclusively in large transmission substations. This has been achieved with a huge amount of hardwiring to a central RTU. The station RTU then communicates to a SCADA system, forming the only level of data communications in a substation monitoring and control system.

Control has now been moved to the bay level with power system automation. Control may be executed by a dedicated bay controller, or may take the form of control function blocks in an IED, alongside the protection, monitoring and measurement function blocks. (The advantages and disadvantages of each are discussed in Chapter 11.) As with protection, control functions will be seen as an application, rather than a device.

Local control is executed on the lower communication level (between the bay level and the process level), and remote control is dependent on the higher communications level (between the bay level and the station/network level).

A. Lower level

The information needed from the field for the control application is mainly limited to status of switchgear (open/close type digital inputs). This is provided by hardwiring the auxiliary contacts of the switchgear to the input terminals of the control device. This information will possibly be provided in future on a communications bus (shared with protection information) to create a 'wireless' switchgear panel. This will mostly be beneficial where the bay controller is located some distance away from the switchgear, e.g. in a high voltage switchyard. In a typical distribution type switchgear panel, where the bay controller is located within cms from the actual switch, hardwiring will still be the most sensible option.

Requirements

Performance	Level required
Speed/data throughput	High
Response time	High
Time synchronization	High
Avalanche handling	High
Data integrity	High
Link availability	High
Data priority	High

Note:
All the requirements are high, as certain local control commands, for example open a circuit-breaker due to the occurrence of a critical event, can have the same priority as a protection-initiated command.

B. Higher Level

This is the communication level from the control AP to the station or central SCADA. The local control and remote control applications are interdependent, with much information shared between them. Information sent from the bay level control AP to the SCADA will include status information, command acknowledgment, and protection lockout information. Data sent from the SCADA to the control AP will include program configuration and specific commands.

Requirements

Performance	Level required
Speed/data throughput	Medium to High 1
Response time	Medium to High 1
Time synchronization	High
Avalanche handling	High
Data integrity	High
Link availability	Medium to High 1
Data priority	High

Explanatory notes:
- The requirements for speed, response time and link availability will be medium to high, depending on the criticality of the data. 'Normal' data will generally be medium, whereas the requirements for critical data, for example in an inter-bay interlocking scheme, will be high.
- The requirement for time synchronization will be high for sequence-of-event reports.
- Avalanche handling, data integrity and data priority will be high as generally all control commands and related data to/from the SCADA must receive first priority in all circumstances, without compromising integrity.

9.3.3 Measurements

Measurements may include electrical measurements, such as voltages, currents, power, power factor and harmonics, as well as other analog values obtained from transducers, e.g. transformer and motor temperatures.

Traditionally, different sets of CTs are used for protection and metering applications. However, IEDs in a power system automation system use only one set of CTs for both applications. Therefore, CTs should be used which are accurate enough for both applications. Most manufacturers' IEDs will guarantee an accuracy of 0.5% or better from 1% to 600% of FLC. If a higher accuracy is required, e.g. for electricity billing purposes, a dedicated metering system should be employed, and the metering information could still be sent to the SCADA through the power system automation communications system. Dedicated metering devices have powerful data storage and processing capabilities, and updated information would be sent through to the SCADA, rather than raw data.

A. Lower level
This is the raw input from the field instruments (e.g. CTs, VTs, temperature transducers, etc) to the measuring application. The same instruments may be shared with the protection application, or separate dedicated metering instruments may be used. The metering device may send its information to the IED or directly to the SCADA (which would normally be the case). Measurement data in the context of power system automation is for information and alarm purposes only, and is not needed for the control or protection applications. When the latter is the case, the protection and/or control applications will obtain the data directly from the field instruments as an input to their function blocks.

Requirements

When the measurement application shares the same field instruments with the protection application, the communication requirements as discussed in section. 9.3.1 will be relevant, for the lower level communication requirements will always be higher for protection than for measurement information.

If separate field instruments and metering equipment were used, the following table would be relevant:

Performance	Level required
Speed/data throughput	Low
Response time	Low
Time synchronization	Low
Avalanche handling	Low
Data integrity	Low
Link availability	Low
Data priority	Low

Note:
All requirements are low, as the data is for information and non-critical alarm purposes only. This does not mean that the data may be discarded. It is still required, but as a lower priority to the other applications. Data errors will be corrected the next time the information is updated, which will usually be within seconds.

B. Higher level

Measurement information will be sent through to the SCADA, either by the IED, or by other dedicated devices capable of communicating on the network. IEDs and other devices will generally have sufficient data storage capacity, so that data can be stored when the communication link to the SCADA is not available, and send when the link becomes available again.

Performance	Level required
Speed/data throughput	Low
Response time	Low
Time synchronization	Low
Avalanche handling	Low
Data integrity	Low
Link availability	Low
Data priority	Low

9.3.4 Monitoring information

Monitoring refers to information such as switchgear status (those not used directly as inputs by the local control application), trip circuit status, condition monitoring information, maintenance information, etc. The urgency of this information is generally considered as low, and can be sent through as soon as the communication link allows it. This information will normally be used to generate alarms and event records (not sequence-of-event reports of important information, which will be part of the protection or control function blocks).

Monitoring information of high priority will usually constitute inputs directly to the control or protection applications.

A. Lower level

This is the information from the switchgear and other monitored devices (e.g. a battery tripping unit) to the monitoring application. Some of the information is assembled by the IED itself, e.g. trip circuit supervision, trip operation counters, etc. Other information is usually in the form of digital inputs to the IED.

Requirements

Performance	Level required
Speed/data throughput	Low
Response time	Low
Time synchronization	Low
Avalanche handling	Low
Data integrity	Low
Link availability	Low
Data priority	Low

B. Higher level

This is the level from the monitoring application to the SCADA.

Requirements

Performance	Level required
Speed/data throughput	Low
Response time	Low
Time synchronization	Low
Avalanche handling	Low
Data integrity	Low
Link availability	Low
Data priority	Low

9.4 Example of Requirements

Table 9.2 is an example of requirements that will be applicable to a typical distribution substation.

Example	Response Time	Data Integrity
Event recordings	1–20 ms	Low
Alarming	1 s	Medium
Command	0.5 s	High
Interlocking BLOCK	<10 ms	High
Interlocking RELEASE	1 s	High
Diff protection TRIP	1–5 ms	High
Parameter set	3 s	Medium
Fault analysis	>1 min	Low

Table 9.2
Example of communication requirements in a substation

10

Power system automation architectures

10.1 Introduction

Different architectures are possible to implement a *power system automation* system, and a multitude of different systems exist today all over the world. However, the first systems employed to obtain control and supervision over substations, which basically consist of a central RTU (to which all the I/Os were hardwired) communicating to a SCADA system, are not considered *power system automation* systems, by definition. These systems do not incorporate local bay intelligence, and are referred to as substation control and supervision (SCS) systems.

Power system automation consists of local intelligence, a communication network accessible to all applications and devices, and at least one SCADA station. More than one SCADA station may form part of the power system automation network. The SCADA station(s) may communicate to a higher level communication network, e.g. an intranet or even the Internet (caution!), but this higher network is not considered part of the *power system automation* system.

Modern systems tend to move towards more common architectures, and four main types of architecture may be identified today, as discussed in this section. The various architectures differ only in the bay level, with the process level, communication network and SCADA level having mainly the same properties in all architectures.

10.2 Types of power system automation architectures

10.2.1 Type 1 systems

This is one of the most advanced systems available today, and will probably be the system employed in most distribution substations in future.

The type 1 system is illustrated in Figure 10.1.

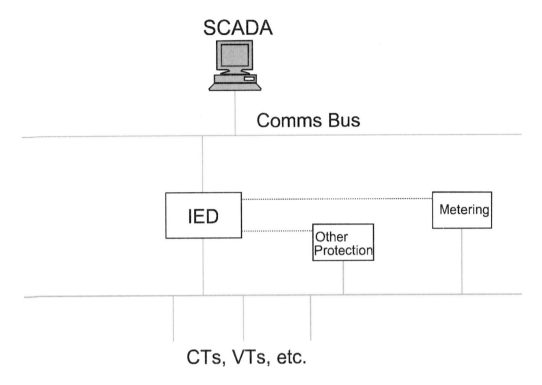

Figure 10.1
Type 1 system

The system consists of one IED handling all the power system automation applications, namely the protection, local control, measurement and monitoring applications. The IED communicates directly to the communication network.

The IED may be assisted by dedicated protection devices (such as line differential protection, for instance) and dedicated metering devices. Although these supplementary devices may communicate directly to the network, it is preferable to communicate to the IED, in order to have an integrated protection and control philosophy. (Otherwise, the IED will not know when, for instance, a supplementary protection device has tripped on an electrical fault, and critical information for the control and sequence-of-event recordings will be lost.)

Advantages of type 1 systems

- Simplified circuitry and wiring, as all I/Os are routed to one device
- Integrated protection and control functions in one device, making programming and configuration of the system easier
- Installation, programming, configuration and commissioning of only one device per bay (not taking supplementary devices into account)
- Integrated disturbance recordings and sequence-of-event reporting, as everything is occurring within one device
- Self-supervision (internal fault diagnostics) covers all applications
- Less loading on the communications network, also resulting in easier configuration of the network, as all bay communications originate from one device
- Easier fault location in case of network communication problems

- Physical space saving on switchgear panels
- Peer-to-peer communication of IEDs possible (depending on communication topology and protocol)

Disadvantages of type 1 systems
- All bay applications will break down in case of total failure of the device
- All data acquisition and supervisory control to all applications are lost in case of communication failure to the device
- Other applications' demand on microprocessor time may marginally slow down execution of the protection functions

Type 1 systems are offered by the following leading manufacturers (see Chapter 12 for a discussion of each manufacturer's system and related products):
- ABB
- GE
- Siemens
- ALSTOM

10.2.2 Type 2 systems

The type 2 system architecture is illustrated in Figure 10.2.

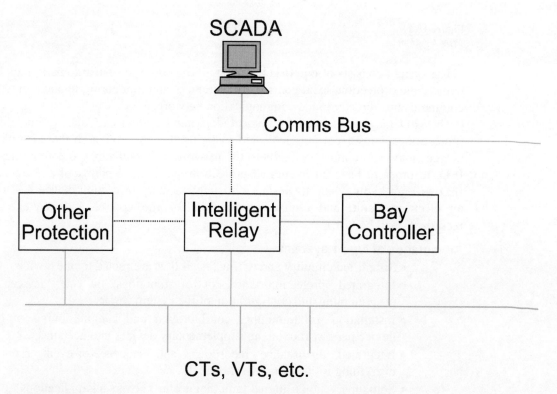

Figure 10.2
Type 2 system

The system differs from a type 1 system mainly in the fact that the two high priority applications, namely protection and control, are separated in two independent devices. The bay controller normally handles the higher level communications to the SCADA network, and a direct communication link is established between the bay controller and protection device. The protection device may or may not communicate directly to the SCADA network.

The protection application will reside in an intelligent protection relay, which will (or should!) have all the functions of the protection application in an IED. Sometimes only a dedicated protection device (e.g. differential protection) will be used for a specific bay, without the more basic protection functions of the typical IED.

The control functions will reside in a bay controller ('mini RTU'), which in turn will typically have all the functions similar to the control application of an IED.

Certain essential information, e.g. for bay interlocking, sequence-of-event reports, etc, may be exchanged with a direct link between the protection relay and bay controller.

The two lower priority applications, namely monitoring and measurement information, will reside in either of the two devices. Normally, the monitoring application will reside in the bay controller, and the measurement application in the protection relay, as this is where these applications are most needed. However, this will depend on the specific manufacturer's product capabilities, and the design philosophy of the system.

Advantages of type 2 systems
- Dedicated control and protection devices. Failure of one device will not influence the applications of the other device
- Execution of protection functions will not be slowed down due to other applications' demand on microprocessor time
- Peer-to-peer communications of bay controllers possible, depending on communication topology and protocol used

Note: (Peer-to-peer communications on a network between the protection relays will normally not be required with a type 2 system. Protection relays will have point-to-point communications in differential protection schemes, or a dedicated network for busbar zone protection.)

Disadvantages of type 2 systems
- More complex circuitry and wiring compared to Type 1 systems
- Protection and control applications not integrated within one device, making programming and configuration more complex
- Installation, programming, configuration and commissioning of at least two devices per bay
- Disturbance recordings and sequence-of-event reporting not integrated in one device, placing higher demands on time synchronization
- More loading on the communications network, also resulting in more complex configuration of the network, as bay communications originate from more than one device
- More physical space required on switchgear panels

The following leading manufacturers offer products that support type 2 systems:
- ABB
- GE
- Siemens
- ALSTOM

10.2.3 Type 3 systems

The type 3 system architecture is illustrated in Figure 10.3.

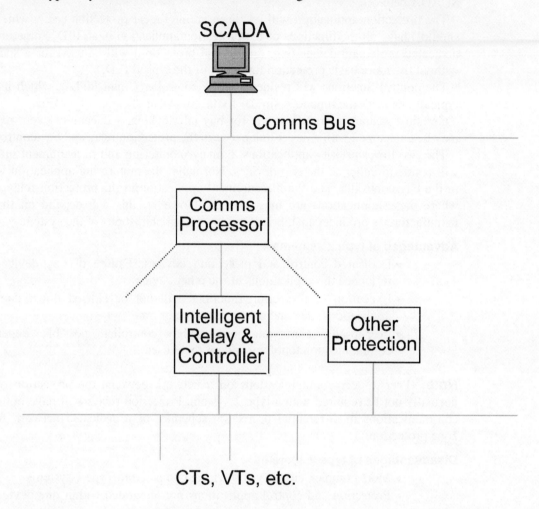

Figure 10.3
Type 3 system

The type 3 system consists of an intelligent relay, which does not communicate directly to the SCADA master, hence it is not termed an IED in the power system automation context. Communications to the SCADA station are handled by a communications processor, which can typically have up to 16 relays connected (hardwired) to it in a star configuration. Other relays may be installed for additional protection functions, but the main, intelligent digital relay will also handle the control, monitoring and measurement applications, in addition to its protection functions. Essentially, this relay does basically everything the type 1 IED does, without communicating directly to the SCADA station.

The communications processor in a type 3 system is designed specifically for the rugged substation environment. The processor is usually able to handle limit control functions via separate I/Os. However, the communications processor's main task is communications to the SCADA station on behalf of the relays. This is what differentiates the comms processor from a RTU or PLC, as in type 4 systems. Type 3 systems have all the main control and monitoring functions embedded in the local (bay) intelligence of the main digital relay.

The comms processor in type 3 systems is normally very versatile and powerful concerning data communications, as this is its main function, and various protocols are usually supported. The comms processor is generally also able to support various types of devices in the substation from many different manufacturers.

Advantages of type 3 systems
- Simplified circuitry and wiring (compared to type 2 and traditional systems), as all I/Os are routed to one device, with only additional wiring between the relays and comms processor
- Integrated protection and control functions in one device, making programming and configuration of the system easier
- Installation, programming, configuration and commissioning of only one device per bay (not taking supplementary devices into account)
- Integrated disturbance recordings and sequence-of-event reporting, as everything is occurring within one device
- Self-supervision (internal fault diagnostics) covers all applications
- Less loading on the communications network, also resulting in easier configuration of the network, as all station communications originate from one device
- The microprocessor does not have to handle high level communications on the network, making more processor time available to the other applications
- Very easy fault location in case of network communication problems
- Physical space saving on switchgear panels

Disadvantages of type 3 systems
- All bay applications will break down in case of total failure of the intelligent relay
- All data acquisition and supervisory control to the whole substation is lost in case of hardware failure or communication failure to the comms processor (or part of the substation if more than one comms processor is employed)
- Other applications' demand on microprocessor time may marginally slow down execution of the protection functions; however, this is improved compared to type 1 systems due to the absence of high level communications
- More complex circuitry and wiring compared to type 1 systems
- Peer-to-peer communications between relays are not possible on a network. Direct point-to-point communications between two relays may be possible

The following leading manufacturers offer products that support type 3 systems:
- SEL *[handwritten: this is why SEL is dedicated to product quality and a long product warranty and outstanding support.]*

10.2.4 Type 4 systems

The type 4 system architecture is illustrated in Figure 10.4.

The type 4 system is close to the type 2 system, but the bay controller and protection relay do not communicate directly to the SCADA station on the higher level communication network. The relay(s) will be more traditional, protection orientated, without control applications, but with the ability to communicate to the bay RTU or PLC.

Local bay control applications will be handled by a bay RTU or PLC, to which the relay will be directly connected. Higher level communications to the SCADA station will generally be via a station RTU.

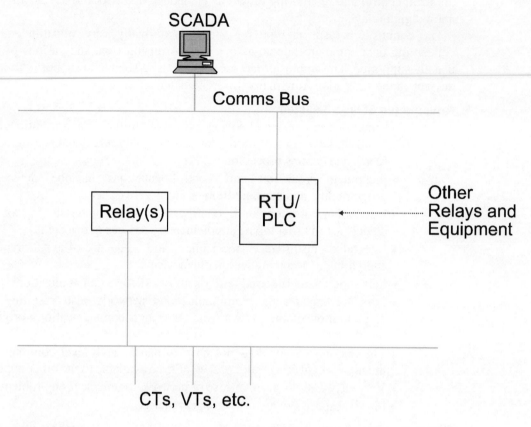

Figure 10.4
Type 4 system

Advantages of type 4 systems

- Dedicated control and protection devices. Failure of one device will not influence the applications of the other device
- Execution of protection functions will not be slowed down due to other applications' demand on microprocessor time
- The microprocessor of the relay does not have to handle high level communications on the network, making more processor time available to the other applications
- Peer-to-peer communications of bay RTUs/PLCs possible, depending on communication topology and protocol used

Note: (Peer-to-peer communications on a network between the protection relays will normally not be required with a type 2 system. Protection relays will have point-to-point communications in differential protection schemes, or a dedicated network for busbar zone protection.)

Disadvantages of type 4 systems

- More complex circuitry and wiring compared to the other type systems
- Protection and control applications not integrated within one device, making programming and configuration more complex
- Installation, programming, configuration and commissioning of at least two devices per bay

- Disturbance recordings and sequence-of-event reporting not integrated in one device, placing higher demands on time synchronization
- More physical space required on switchgear panels
- All data acquisition and supervisory control to all applications is lost in case of communication failure to the bay RTU/PLC, or in case of hardware failure of this device

The following leading manufacturers offer products that support type 4 systems:
- Siemens
- Various RTU and PLC suppliers

11

Power system automation systems on the market

11.1 Introduction

The following discussion is by no means intended to be exhaustive of what is available in the market regarding power system automation at present, but only highlights some of the leading manufacturers' systems and products as examples of what can be done today, and where the technology is heading.

Systems on offer from GE, ABB, SEL, Siemens and ALSTOM are briefly examined.

11.2 GE

11.2.1 Overview of products

The *power system automation* system offered by GE centers around the universal family of relays (UR range). GE predominantly uses the term 'relay' and uses the term IED very tentatively in their literature, but the UR range really conform to the description of an IED in the context of power system automation, incorporating protection, control, metering, monitoring and communications in one device. The architecture offered by GE is a typical example of a type 1 configuration. The configuration is illustrated in Figure 11.1.

As can be seen, the IED communicate directly to the SCADA system on a common communication bus. Peer-to-peer communications are supported depending on the protocol used.

The UR family of IEDs is built on a common platform, with different protection features in each member of the family.

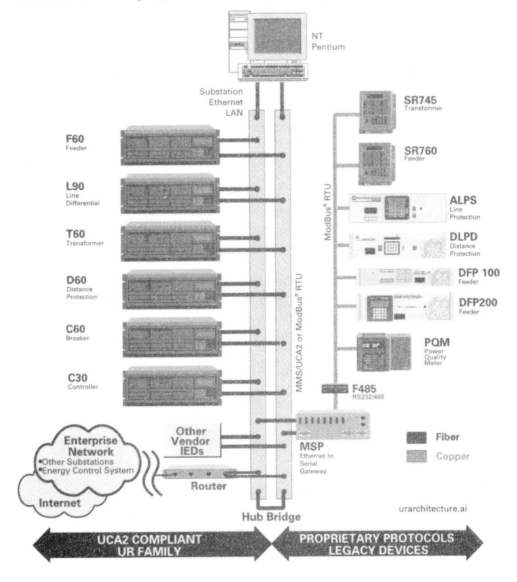

Figure 11.1
GE's proposed SA architecture

(Source: GE Power Management Products CD)

11.2.2 Protection

The protection features in each device are what differentiate the members of the UR family from each other. At present, there are seven members of the family, as follows:

L90	: Line Differential Protection
L60	: Line Phase Comparison Protection
D60	: Distance Protection
T60	: Transformer Protection
F60	: Feeder Protection
C60	: Breaker Protection
C30	: Controller (typical bay controller without any protection functions)

- L90: Current differential, with overcurrent backup protection. High-speed tripping of less than one cycle can be obtained with the differential protection, and a fault location function is supported by the monitoring application.
- L60: This protection application uses the comparison of current phases to detect fault conditions and is designed for HV and EHV transmission lines. Overcurrent and ground fault protection are included as backup, with a sensitive current disturbance (fault) detector.
- D60: High-speed distance protection for MV and HV transmission lines. Four zones of distance protection are included, with overcurrent and ground fault backup, as well as sensitive overcurrent detection and under-voltage protection. Fifteen different current–time characteristics can be selected, including two user-defined curves.

The following curves are provided:

IEEE	Extremely inverse
	Very inverse
	Moderately inverse
IEC	Curve A (BS142)
	Curve B (BS142)
	Curve C (BS142)
	Short Inverse
GE IAC	Extremely inverse
	Very inverse
	Inverse
	Short Inverse
	I2t
	Definite Time
Custom	Flexcurve A *
	Flexcurve B

A FlexCurve is constructed by superimposing different curve shapes, as illustrated in Figure 11.2. (The term 'FlexCurve' is a trademark of GE.)

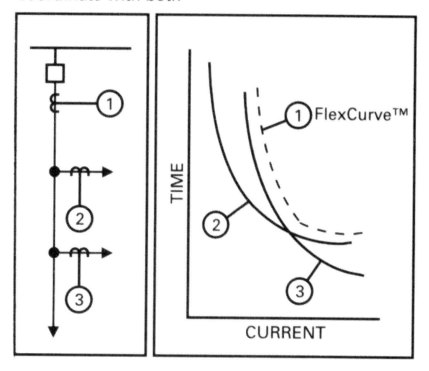

Figure 11.2
Example of the use of a FlexCurve ™

(Source: GE Power Management Products CD)

- T60: Transformer protection for small, medium and large power transformers, including differential protection with harmonic restraint. Multiple overcurrent elements are included.
- F60: Feeder management relay for feeder protection and metering. Full three-phase, neutral and ground fault overcurrent, time delayed and instantaneous. Fifteen time–current curve shapes provided, as above. Sensitive ground fault and under-voltage protection included.
- C60: Breaker management relay for breaker control, monitoring and protection. This IED includes complete breaker supervision, synchrocheck, and autoreclosure functions. Overcurrent, under-voltage, underfrequency, and sensitive ground fault protection are included.
- C30: Bay controller without any protection or metering functions.

11.2.3 Control

All IEDs in the GE UR family offer local and remote control facilities. The IED is programmed by the use of logic equations (FlexLogic, a trademark of GE).

The basic configuration architecture is illustrated in Figure 11.3.
The C60 relay includes load shedding, autoreclosure and breaker failure functions.

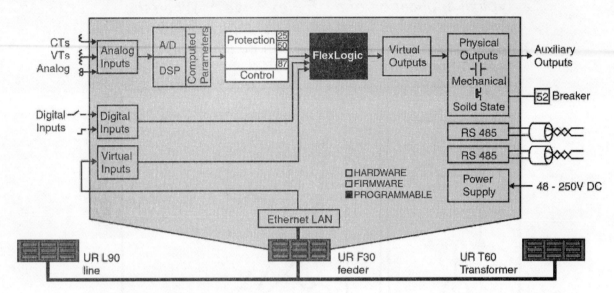

Figure 11.3
UR family basic configuration architecture

(Source: GE Power Management Products CD)

11.2.4 Metering

All IEDs include current, voltage, power and frequency measurements, including current and voltage phasors.

11.2.5 Monitoring

All IEDs feature:

- Disturbance records: 1×128 to 31×8 cycles (configurable, pre- and post-event); 64 samples per cycle
- Event records: Up to 1024 events, time-tag 1 microsecond
- Self-diagnostics

The T60 features tap position monitoring and % of rated load. The C60 features trip circuit supervision, breaker operations counter, and breaker contact arcing current monitoring.

11.2.6 Communications

An RS-232 port on the front of the unit is intended for local interrogation and programming using the URPC program (see next section) on a standard 19.2 kbps ModBus RTU protocol. A standard RS-485 port is provided on the rear for the ModBus RTU protocol, supporting a baud rate of up to 115 kbps. A second port on the rear can be either another RS-485 port or a 10 Mbps Ethernet port, supporting MMS/UCA2 and ModBus TCP/IP protocols.

11.2.7 SCADA

A Windows based proprietary software program, called URPC, may be run on a PC with the Windows 95/98/NT operating systems. The program may be used locally on the front RS-232 port, or remotely on the rear ports. Full access to relay real and historical data is provided.

11.3 ABB

11.3.1 Overview of products

ABB offer various products for SA, depending on the required architecture. Products are available to setup either a type 1 or type 2 system, as discussed in Chapter 10.

11.3.2 Type 1 products

The cornerstone of the type 1 system from ABB is the REF series of relays. The REF relays were developed as feeder protection relays, and ABB has duped them 'feeder terminals'. The series consists of the REF 541, REF 543 and REF 545 relays. The only difference between the relays in the series is the amount of I/O each relay can accommodate. The REF relays are IEDs in every sense of the word; offering advanced protection, control, monitoring, metering and communication capabilities. The REF relays have been designed for medium voltage distribution type networks.

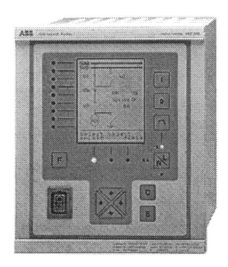

Figure 11.4:
ABB REF 54_ relay

Protection

The REF series offer the following protection function blocks as standard features in the protection application. The function blocks can then be either configured or de-activated, depending on the functions required.

- Non-directional overcurrent (low-set, with seven different curve types, high-set and instantaneous)

- Non-directional earth fault (low-set, high-set and instantaneous)
- Directional overcurrent (low-set, high-set and instantaneous)
- Directional earth fault (low-set, high-set and instantaneous)
- Phase discontinuity
- Three-phase over-voltage (low-set and high-set)
- Residual over-voltage (low-set, high-set and instantaneous)
- Under-voltage (low-set and high-set)
- Under/over-frequency (5 stages)
- Thermal overload
- Transformer inrush and motor startup current detector
- Autoreclosure function
- Synchrocheck function

Control functions
The control functions include local and remote control of up to six switching objects (circuit-breakers, disconnectors, or earthing switches), status indication of the switching objects, and interlocking on bay and station level. Information of alarm channels form part of the control application.

The indication of objects can be configured on the local graphic display. The mimic configuration can be designed by the user.

Measurement
Measurement functions include three-phase currents, neutral current, three-phase voltages, residual voltage, frequency, active and reactive power, reverse power, power factor and harmonics.

The transient disturbance recorder forms part of the measurement application. The recorder is used for monitoring the current and voltage waveforms as well as status data of logic signals and binary inputs connected to the relay terminals. Up to 16 analog and digital channels may be monitored. The sample rate is 40 samples per fundamental cycle.

The recording length depends on the number of records and channels used. For example, 1066 cycles may be recorded for one record on one channel, and 12 cycles per record for 10 records on 10 channels each.

The recording can be triggered by one or several of the following occurrences:
- Rising or falling edge of any or several of the binary inputs
- Overcurrent, over-voltage or under-voltage
- Manual triggering via MMI
- Serial communication parameter
- Periodic triggering

Monitoring
Monitoring functions include the condition monitoring functions like trip circuit supervision, operation time counter, circuit-breaker electric wear, breaker travel time, scheduled information, and self diagnostics. Other functions include auxiliary voltage and internal overtemperature indication. Gas density monitoring is available for SF6 switchgear.

Communication
An optical RS-232 optical connection is available on the front panel for ASCII type communication to a local PC or laptop. This is intended for initial configuration during commissioning, but may also be used for local download purposes. A rear RS-485

connection, with an optional fiber optic interface, is available for connection to the network communications. Two protocols are supported as a standard, namely the ABB SPA-bus protocol (up to 19.2 kbps) and LON bus (up to 1.25 Mbps). LON bus is synonym to the LonTalk protocol developed by Echelon Corporation. Peer-to-peer communications are only supported by LON bus.

The IEC 870-5-103 protocol will also be supported.

SCADA

The REF relays are designed to communicate via LON bus to ABB's MicroSCADA control system. A maximum of 255 subnets and 127 nodes per subnet is supported on one SCADA system. More than one SCADA system may be connected in a LAN/WAN network.

11.3.3 Type 2 products

ABB's SPA series of devices offer various methods of implementing type 2 systems. A comprehensive range of protection relays, bay controllers and IEDs is available. This series of equipment is designed to communicate via ABB's SPA Bus to the SMS 010 Substation monitoring system.

11.4 SEL

11.4.1 Overview of products

SEL's *power system automation* system centers around intelligent relays and the communications processor, and is a typical example of a type 3 system as described in Chapter 10.

Various different types of relays may be utilized for different protection functions. The type SEL-351 relay is the premier multi-function relay for transmission and distribution applications and provides comprehensive protection, control, metering, and monitoring applications, all in the one device. Up to 16 relays can be connected to the communications processor, type SEL-2020 or SEL-2030. The SEL-2030 comms processor differs from the SEL-2020 in that it is readily adaptable to new communication protocols. When a new protocol needs to be added to the SEL-2030, a protocol card can simply be slotted into the appropriate space.

The comms processor regulates all communication in the substation between itself and the relays, and all higher level communications to the SCADA network.

Figure 11.5 is an example system, as published by SEL on their website:

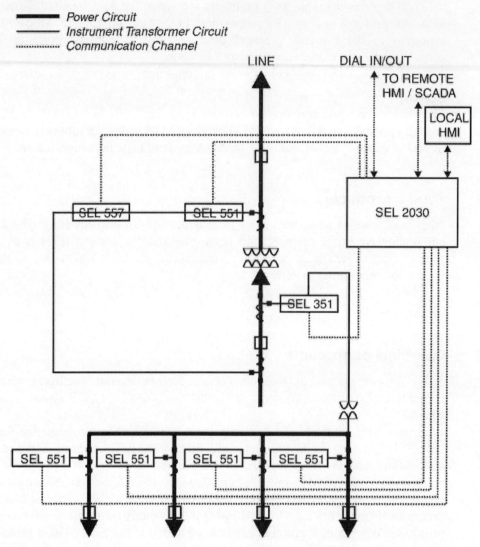

Figure 11.5
SEL example system

Peer-to-peer communications on a multidrop bus network is not supported, as SEL uses a star topology. Instead, SEL has developed their MIRRORED BITS™ technology, whereby fast point-to-point communications are supported over a direct fiber optic link. Initially, this was only available for relay-to-relay communication between two relays, but they have subsequently developed the SEL-2100 protection logic processor, supporting up to 15 devices.

11.4.2 Protection

SEL microprocessor-based relays, all compatible with the SEL comms processors, are available for virtually every protection application in the generation, transmission, distribution and industrial markets.

They tend to specify their relays according to functions rather than application, which can be quite restrictive. For example, the SEL-387E relay is specified as a 'Current differential and voltage relay'; whereas this relay also provides transformer protection, overexcitation-, over- and underfrequency protection, as well as metering, monitoring and automation functions.

The result is that one has to examine the relay data sheets quite thoroughly to locate the right relay for the right application. SEL offers a free software program to assist the user to find the most suitable relay for the required application. The user would key in his requirements and the program would suggest which relay(s) to use.

The basis of SEL's power system automation system is the multi-function SEL-351 relay.

The SEL-351 'Multifunctional directional overcurrent' relay offers the following protection functions:

- Phase and ground directional overcurrent
- Non-directional phase and ground overcurrent
- Under- and overfrequency (six steps)
- Under- and over-voltage
- Synch check
- Autorecloser
- Reverse power
- Breaker failure detection

11.4.3 Control

- Sequential events recorder, automatically transfers data to SEL-2030
- Up to 16 remote control devices per relay

11.4.4 Metering

- Accurate metering of all electrical values, eliminate need for panel meters
- Load profiling of up to 15 user-selected values, including V, A, f, W, VAR, Wh, VARh, PF, peak demand
- Stores up to 40 days of data for a load profile of 15 values logged every 15 minutes
- Disturbance reports for pre- and post-fault analysis; 15- or 30-cycle selectable
- Voltage sag/swell/interruption reports, event triggered and time-tagged

11.4.5 Monitoring

- Substation battery voltage level monitoring
- Intelligent breaker monitoring
- Status monitoring of up to 16 devices
- Self supervision

11.4.6 Data communications

SEL uses an internally developed protocol, collectively termed *fast meter protocol*, for communications between their relays and the comms processor. This is an interleaved ASCII type data communication.

Serial communications to other manufacturers' products are supported by the SEL-2030.

A multitude of higher level communications to the SCADA network can be supported by the SEL-2030. Currently, the following protocols are supported as a standard:
- MODBUS
- DNP V3.0
- IEC-870-5-101
- UCA

SEL claims that virtually any other widely used protocol can be supported by the SEL-2030 on request.

The SEL-2030 can support communications on all active ports (up to sixteen) simultaneously. Additional to communication control, the SEL-2030 also perform data storage and processing, auto-configuration of IEDs, time synchronization, and SELOGIC® control equations (for example transmitting a certain message when an event occurs).

Fast point-to-point communication (less than 10 ms) can be achieved between two relays using SEL's MIRRORED BITS technology.

SEL developed the SEL-2100 protection logic processor, which can accommodate up to 15 remote devices, communicating via MIRRORED BITS. This processor also incorporates advanced local intelligence to coordinate complex trip and close decisions. A typical application would be to create a very efficient and low cost busbar protection scheme, using the same relays installed for feeder protection (for busbars of less than 15 circuit-breakers).

11.4.7 SCADA

SEL do not have their own SCADA software. Instead, they would typically rely on a SCADA software vendor or system integrator to be involved to establish the SCADA part of the power system automation system.

Preferably, the SCADA software should support one of the protocols already supported as a standard by the SEL-2030; if not, a protocol card can be added to the comms processor to support the protocol(s) preferred by the software or the end-user.

11.4.8 Supplier's information

The following are the most important advantages proposed by SEL of their system:
- The SEL communications processor can communicate directly to virtually all microprocessor-based relays. Devices from different manufacturers with different protocols, including protocols designed for multidrop applications, are supported on the star network, as each has a dedicated direct connection.
- The SEL star network design supports a wide range of IED capabilities. Simple, slow communicating devices can coexist with more complex fast communicating relays on the same system.
- The SEL star network is a truly open architecture and will accommodate multiple protocols, baud rates, and network interfaces.
- Direct connections in a star network eliminate data overhead in the form of addressing, increasing data throughput.
- Troubleshooting communication problems is quick and easy with a star network. Easy verification is supported on the SEL comms processor in the form of LED indication.

- Single relay communication failure will not influence the network, as for example a relay failing to release communication control in a multidrop network.
- The comms processor simplifies implementation through autoconfiguration. This process automatically determines the proper baud rate to communicate with the connected relay as well as startup parameters, device type, and capabilities.
- The comms processor can eavesdrop on conversations between two devices, capture, and store the data.
- The resources in the relay can be focused on optimizing protection solutions.
- New protocols can be easily added to the comms processor. A single device needs to be upgraded when protocol requirements change, instead of each of the relays individually. This is more economical and the relays are left undisturbed and in service as a protocol change is made.
- The comms processor's capability to communicate to different relays using different protocols, extends the useful lifetime of the relay, as the relay does not have to support a new protocol.
- New and future IEDs in the substation are easily accommodated and integrated.

The following is a complete list of advantages as published on their website:

Advantages with star network (SEL)

- A star network allows the migration of *some* of the communications functions from the IED to the network controller. Moving protocols into the IEDs adds to their cost and accelerates their obsolescence as technology advances. The resources available within the IED are instead better focused on optimizing protection solutions.
- A communications processor used as a network controller can act as a client/server, data concentrator, substation archive, programmable logic platform, gateway, router, dial-out device, communication switch, and time synchronization broadcaster.
- Time to perform control action can be faster from master to IED than in bus topology.
- The communications processor can communicate without developing vendor-specific protocol software and can eavesdrop on conversations between two devices in the I&C system.
- Star networks can acquire and transfer data using much slower direct connections since it is performing many conversations simultaneously. These direct connections are also more reliable, more robust, and less expensive.
- The communications processor simplifies implementation through autoconfiguration, which describes the relay's attributes and capabilities.
- Direct connections in a star network are all independent and allow the network to support a wide range of relay capabilities. Simple, slow communicating devices can coexist with more complex fast communicating relays.
- The star network is the only design that is truly open and accommodates multiple protocols, multiple baud rates, and multiple network interfaces.

- Communications processors acting as a network controller enhance the value of the I&C system data by making it available to multiple master systems and other users.
- The communications processor creates an autonomous coordinated star network I&C system within the substation that does not rely on a master connection.
- Star networks allow mediation of local or remote control of the entire substation.
- As protocol requirements change in the substation, a communications processor network controller can be upgraded instead of each of the relays individually. The relays are left undisturbed and in service as a protocol change is made. It is also more economical to make this change in a single device.
- The age of IEDs that are in substations today varies widely. Many of these IEDs are still useful but lack the most recent protocols. Rarely is a substation integration upgrade project undertaken where all existing IEDs are discarded. A communications processor that can communicate with each IED via a unique baud rate and protocol can extend the usefulness of IEDs. Using a communications processor for substation integration also easily accommodates future IEDs.
- Star network interleaved data streams are a simple innovative way that multiple conversations can occur simultaneously. Multipurpose, historical, and time synchronization conversations can simultaneously occur on a single communication channel.
- Troubleshooting communications problems is much faster and more efficient through simple LED indication on direct links in a star network than attempting to decipher multidrop networks.

SEL admits to the following disadvantages of their system:
- The concept of their star network is not well understood and considered as low tech.
- The user must purchase a network controller regardless of quantity of IEDs. SEL star networks require a communications processor for a single relay installation if the relay does not support the required protocol.
- Addition of nodes requires new point-to-point cable to be installed.

Future development

The following is an extract from SEL's website regarding their future development plans. Note that they regard data communication capability development as a cornerstone of their product development.

'As part of SEL's development plans, we continue to identify the need to implement non-SEL standard protocols within SEL products to support loose relay sales and multidrop system designs.'

'Serial'

'We continue to implement non-SEL serial protocol interfaces within SEL products. The three most popular non-SEL protocols, are DNP V3.0, ModBus, and IEC 870-5.'

'Ethernet'

'We are developing a modular Ethernet interface in the form of the SEL-2701 Ethernet Processor Card. This card will interface to the SEL-2030 Communications Processor, and therefore to relays connected to the SEL-2030, as well as future platforms. The first release of the SEL-2701 will support telnet, a virtual terminal application, ftp, file transfer protocol and UCA/MMS protocol via IO and 100 Mbit Ethernet connections. Customers intend on using these telnet and ftp protocols to transfer settings data and historical data such as event reports and profile data. Some customers are even using ftp to transfer supervisory and control data.'

'The SEL strategy to implement UCA has been to build onto our success with fast, cost effective and reliable substation integration rather than obsolete all existing IEDs. We plan to add UCA connectivity to completely integrated in-service and new substations at the network controller level and then migrate UCA into individual relays and controllers. Our customers can utilize their installed base of IEDs and their existing communications architecture as they phase in UCA. This strategy allows customers to migrate to UCA with minimal disruption and expense. It also gives instantaneous UCA interconnection to most IEDs available or installed today since the SEL-2030 is compatible with all SEL relays and controllers and many other vendor products.'

'Recognizing the market will wish to have UCA connectivity direct to the relays and controllers, future relays will support a direct UCA connection. The SEL 2020/2030s assure backward compatibility to every SEL IED ever sold as well many other vendors' IEDs. Previously installed and new IEDs with simple EIA-232 interfaces are easily connected to the SEL-2020/2030. The SEL-2030 can then bridge these IEDs to other network technology including ModBus Plus, UCA/MMS over Ethernet, and DNP V3.0 over Ethernet.'

'Using the SEL-2030 with embedded UCA technology as the substation controller and future relays with embedded UCA technology, a hybrid system can be created that designs itself as individual devices are chosen based on their merit rather than their interface. Customers will be able to choose from the best in-service or new IEDs to perform control, monitoring, automation, protection, analysis, tests, maintenance, and operation of the power system.'

'We are also committed to develop DNP V3.0 over Ethernet. SEL believes this protocol has many advantages, and it is popular with SCADA vendors. Several of our customers who are actively designing SCADA systems are using this interface. The SEL-2030 supports the traditional serial connections and with the SEL-2701 will support simultaneous connections to UCA/MMS, ftp, telnet, and DNP V3.0 over Ethernet.'

'SEL-2701 and utility communications architecture'

'We committed to the Ethernet because our independent engineering choice was to use Ethernet and TCP/IP protocol, since it is fast, and because of its broad installed base there are many Internet and intranet tools available to enhance development and applications.'

'UCA requires that an application layer protocol, such as MMS, run on top of Ethernet and uses data modelling as described by the UCA compatibility specifications.'

'We have defined the SEL-2701 and the host shared memory interface so that we will be successful in creating sufficient UCA models both in future products and for devices connected to the SEL-2030.'

11.5 Siemens

11.5.1 Products

Siemens offers two main systems, namely the SINAUT LSA system, which follows the more traditional approach and is aimed more at the large transmission type substations, and their new SICAM technology, which approaches the other local intelligence focused systems as discussed in this chapter. The SICAM concept consists of SICAM PCC, SICAM SAS and SICAM RTU.

Siemens follows a very modular approach in their systems. The digital protection relays are generally focused exclusively on protection functions, with added (limited) communication capabilities. Control, monitoring, as well as metering functions are performed by the other devices used in the various systems, namely station controllers, RTUs, and/or bay controllers.

An integrated control and protection philosophy is not very prominent in Siemens' systems. They rather promote a lesser distinction between substation control and process control. Their systems only vaguely correspond with the definition of power system automation, as defined in this text.

The same protection relays are generally used in the different systems, hence protection will first be discussed briefly, and thereafter the different control systems.

11.5.2 Protection

Siemens has a full range of protection relays from high voltage to low voltage applications. The 7-series (SIPROTEC) relays are generally applicable for high- and medium-voltage, and are termed numerical relays. Except for their generator protection relay, they classify their relays according to function and not application. For example, the 7RW600 relay is termed 'Numerical Voltage, Frequency and Overexcitation Protection', which is applied for the protection of electrical machines.

The relays offer focused protection functions. Communications are supported, either to a PC via an RS-232 port or modem, or to a SCADA bus, generally used in conjunction with I/O units.

Control, monitoring and metering capabilities are generally not included in the relay.

11.5.3 SINAUT LSA

The SINAUT LSA is Siemens' traditional Substation Control and Protection System (Siemens terminology). The concept is aimed at large transmission type networks. Control and protection is not very integrated. Although the higher end of the system still centers around a large station RTU-type device, more local intelligence has been introduced. The term 'Substation Control and Protection System' is more fitting than 'power system automation' system to describe the technology, but it still fits into the definition of the latter term (although only just).

The SINAUT LSA system is a typical type 4 system architecture (refer to Chapter 10), and an example is illustrated in Figure 11.6.

Power system automation systems on the market 151

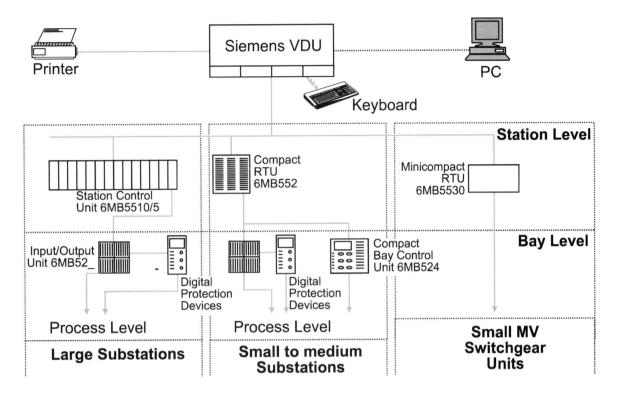

Figure 11.6
Siemens' SINAUT LSA system

Herewith follows a brief functional description of the main devices used in the example. Several other devices are available for use in the SINAUT LSA system. However, most of them will have more limited functionality than those described below, hence the following discussion should be sufficient to demonstrate the basic concept of the system.

Station control unit 6MB5510/5515
- Communications
  The unit functions as the interface between the central control station and the substation for medium to large sized substations. It interfaces to stand-alone I/O devices, master control units and numerical protection devices, and also acts as a communication node.

  A maximum of 121 I/O modules are supported, and a maximum of 3872 items of process information (status indications, measured values, commands, etc). [Note Siemens' use of pre-dominantly PLC-type spoken language.]

  Standard protocols supported are: IEC 870-5-101 and SINAUT 8FW.

  Other protocols may be supported by adding optional modules.

- Control, monitoring and data acquisition
  Operator control and monitoring from one or two central control stations is possible. Sequential and logic functions and interlocking are supported.

Logging, processing and mass storage for archiving of status indications, metered values and protection data. Time synchronization by means of time signal input module.

Compact remote terminal unit 6MB552

The compact RTU performs similar functions to the station control unit, only on a much smaller scale (except for logging and storage of data). Therefore the unit is suitable for medium scope of process signals (i.e. small to medium sized substations).

It can interface to up to seven I/O devices, numerical relays and/or bay control units (maximum of five I/O devices). A maximum of 128 items of process information items is allowed (status indications, measured values, commands, etc).

Sequential and logic functions, interlocking, metering value processing, and time synchronization functions are available.

Up to two SCADA interfaces are supported, with the two standard protocols IEC 870-5-101 or SINAUT 8FW.

Minicompact remote terminal unit 6MB5530-0

This mini-RTU is intended for small MV switchgear units, typically ring main units.

The basic unit provides for 8 indication inputs and 8 command outputs. It can be expanded to a total of 32 indication inputs and 8 command outputs, or 24 indication inputs, 8 measured inputs and 8 command outputs.

The unit is cascadable with up to 3 slave units, maximum distance 500 m.

Indication and measured value acquisition, preprocessing, and telecontrol transmission are performed. Time synchronization is supported, and the two standard protocols IEC 870-5-101 and SINAUT 8FW.

Input/output unit 6MB520

The input/output units are employed in conjunction with the station control unit or compact RTUs. These units acquire and process data and output commands, decentralizing interfacing of signaling contacts, control stations, current and voltage transformers, transducers and numerical protection devices.

Acquisition cycles are adjustable between 10 ms and 1000 ms, time stamped with 10 ms resolution. RMS value computation and smoothing for current and voltage are performed by the unit.

The 6MB520 is used for medium to large substations, and the 6MB522 and -523 for smaller substations, with reduced functionality.

Compact bay control unit 6MB524

The bay control unit is used for decentralized control. It offers local menu-assisted operation via a graphic display, and local intelligent control of switching devices, including device diagnosis and indication lists.

Measured value processing includes instantaneous value recording at sampling intervals of 1 ms, computation of RMS values, active and reactive power, power factor, etc.

Switchgear interlocking and synchrocheck functions are included.

SCADA

The standard SCADA master station, termed LSA control center, consists of a manufacturer-specific VDU (visual display unit), which is used for display and logging of all process data, a functional keyboard for execution of switching actions, and a printer for recording events in hard copy.

An evaluator PC may be connected locally to the VDU, or remotely via a modem. This PC is used for evaluation of the archived data, including events, protection indications, measured values, and fault recordings.

11.5.4 SICAM PCC

The SICAM PCC system is based on a PC control system (PCC stands for PC Controller). The SCADA software runs on a standard commercial PC under the Windows operating system. The system is illustrated in Figure 11.7.

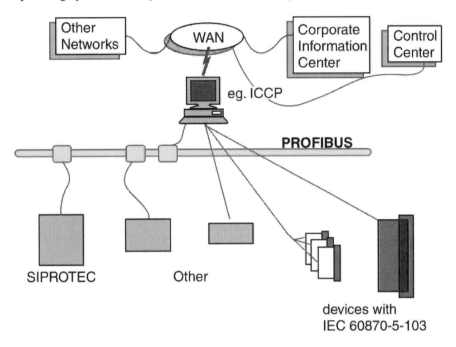

Figure 11.7
Siemens' SICAM PCC system

The system uses PROFIBUS and IEC 60870-5-103 for field communications, and ICCP (inter control center protocol), also known as IEC 60870-6-TASE.2, for communications to a LAN or WAN.

Therefore, any device that can support PROFIBUS or IEC 60870-5-103 can theoretically be integrated into the system.

The PC utilizes the software SICAM WinCC for the field control, which supports remote control, graphical data display, archiving, and evaluation. Interfacing to the LAN/WAN is supported by the software SICAM NET.

SICAM PCC can be regarded as a type 2 system, although it is not a very integrated protection and control solution. The philosophy is basically to provide a common platform of two communication protocols for the field devices, and any compatible device may be hooked up independently.

Siemens recommends the SICAM PCC system for small substations.

11.5.5 SICAM SAS

The SICAM SAS system is recommended for larger substations. SICAM PCC can be expanded and turned into SICAM SAS by adding the station control unit.

The system is illustrated in Figure 11.8.

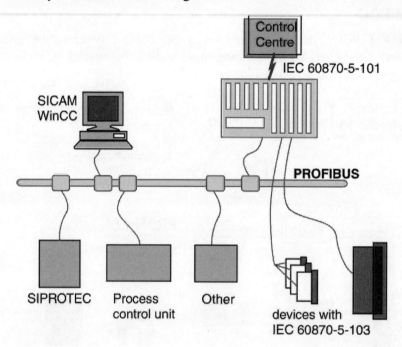

Figure 11.8
Siemens' SICAM SAS system

PROFIBUS and IEC 60870-5-103 are used for field communications, and IEC 60870-5-101 between the station control unit and the control center. Data transmission is event-initiated, and priority of commands is supported.

The control center is a PC, running on a Windows platform, with SICAM WinCC, SICAM SAS, SICAM NET and SICAM plus TOOLS software.

A modular approach is also followed, with the philosophy of establishing an 'open' platform for any compatible device.

SICAM SAS is typical of the type 4 system, as discussed in Chapter 10.

11.5.6 SICAM RTU

The SICAM RTU system has been developed primarily for geographically decentralized substations. The philosophy is based on more centralized control in the RTU, which has powerful PLC type functions. (The functionality is based on the SIMATIC 400 Power PLC.)

The RTU offers stronger communication capabilities to the remote control center. PROFIBUS, Industrial Ethernet, IEC 60870-5-101, and SINAUT 8FW are supported, which means the SICAM RTU may be interfaced to the Siemens SINAUT LSA system.

PROFIBUS are available for field communications.

The system is illustrated in Figure 11.9.

The SICAM RTU system does not strongly support local intelligence. The system is somewhere between the type 4 power system automation system, as described in Chapter 10, and a more traditional substation control system.

Power system automation systems on the market 155

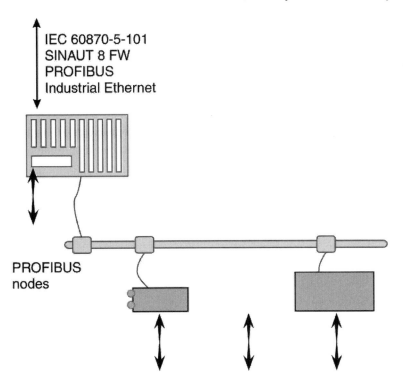

Figure 11.9
Siemens' SICAM RTU system

11.5.7 Suppliers' information

According to Siemens, the distinctions between industrial control and substation control are disappearing, thanks to standards such as Windows, Ethernet and PROFIBUS. They see a future merging of the two worlds of industrial and substation control, and their new SICAM technology as the way to achieve it. That is why the SICAM systems support process control devices on the same bus as protection devices.

The following are the advantages of their SICAM systems, as proposed by Siemens on their website.

Seven advantages of SICAM PCC
- Only standard components are used – resulting in lower implementation cost and training costs.
- The software conforms to Windows standards. Users who are acquainted with modern PC software can use the system intuitively.
- Communications is realized through standardized LANs and WANs, which make use of existing communications infrastructure and dispensing with the need for new investments.
- Ideal for small to medium process volumes.
- The modular architecture supports growth according to requirements.
- The software offers standard integration. Gathered data can be evaluated with the ease of using standard software.
- A number of additional functions are available. The functionality can be extended to suit individual needs.

Seven advantages of SICAM SAS

- Uniform computing, protecting and controlling. All substation control functions featuring distributed intelligence and information processes are based on identical standards to exploit synergistic effects.
- The system is based on international standards, ensuring everything fits, saving effort and protecting your investments.
- The system is open to integration for new innovations on the market.
- All information from the field is stored in one common database.
- Configuration is done by working with function blocks.
- Parameterizing instead of programming. Configure by the click of the mouse or by Drag & Drop style commands.
- SIMATIC® technology included.

11.6 ALSTOM

11.6.1 Product overview

Alstom offers four systems in the field of power system automation, namely the 'S10 Protection Based Substation Control System,' the 'S100 Integrated Substation Protection and Control System,' the 'PSCN 3020 Integrated Protection and Distributed Digital Control System,' and the 'SPACE 2000 Protection and Control Solution for Transmission Substations.'

Alstom offer a multitude of different numerical relays, ranging from very protection-focused relays with limited communication capabilities, to versatile IEDs, *termed protection and bay units by Alstom*. Bay units for control and monitoring are also available to use in conjunction with protection-focused relays.

Measurement and disturbance recordings are generally included in the protection relay, as well as the bay unit, capabilities.

Two types of device, which are the most supportive of power system automation, will briefly be discussed, and thereafter the different power system automation systems.

11.6.2 Protection and bay units

Two devices of note are the 'PS 982 Numerical time-overcurrent protection and bay unit for control and monitoring'; and the 'PD 932 Numerical distance protection with supplementary functions and bay unit for control and monitoring'.

Both devices are based on the same platform, and offer the following control, metering and monitoring functions:

- Control and monitoring of up to 6 electrically operated switchgear units
- Selection from over 200 pre-defined bay types
- Bay interlocking
- Local control and graphic display
- Extensive self-monitoring
- Time-tagged data recording of all electrical values
- Overload data recording
- Ground fault data recording
- Detail measured fault recordings
- Programmable logic

The following communication protocols are supported:
- ModBus
- IEC 60870-5-103, or
- IEC 60870-5-101.

The **PS 982** offers the following protection functions:
- Definite time overcurrent, 3 stages, phase selective
- Inverse time overcurrent, single stage, phase selective
- Short circuit direction determination
- Autoreclosing
- Motor protection
- Unbalance protection
- Over-voltage and under-voltage protection
- Switch on to fault protection
- Breaker failure
- Ground fault detection

The **PD 932** offers the following protection functions:
- Distance protection, four distance stages, six timer stages
- Under-impedance fault detection logic with load blinding
- Definite time overcurrent, 4 stages, phase selective
- Inverse time overcurrent, single stage, phase selective
- Autoreclosing
- Under-voltage protection
- Switch on to fault protection
- Breaker failure
- Ground fault detection
- Thermal overload

11.6.3 Bay units

The *CF 242 Bay Unit* is designed for the control and monitoring of up to 9 electrically operated switchgear units, to be installed in the bays of medium voltage and non-complex high voltage substations. The *CF 202 Compact Bay Unit* is available in three models, for control and monitoring of 3, 5, and 7 units respectively.

The same functionality as the protection and bay units is incorporated regarding control, monitoring and measurement functions; only without the protection functions.

- Control and monitoring of up to x electrically operated switchgear units
- Selection from over 200 pre-defined bay types
- Bay interlocking
- Local control and graphic display
- Extensive self-monitoring
- Time-tagged data recording of all electrical values
- Overload data recording
- Ground fault data recording
- Detail measured fault recordings
- Programmable logic

The following communication protocols are supported:
- ModBus
- IEC 60870-5-103, or
- IEC 60870-5-101.

11.6.4 MiCOM S10

The MiCOM S10 is termed a 'protection-based substation control system'. The system is aimed at industrial and distribution substations, with local supervisory control.

Figure 11.10 illustrates the system.

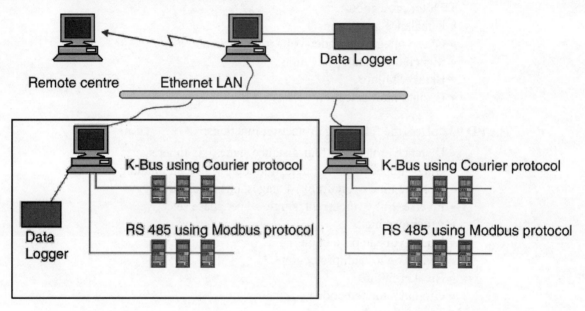

Figure 11.10
ALSTOM MiCOM S10 system

The main substation control station is an industrial or commercial PC with a Windows NT operating platform. Control and monitoring of the system are done via graphical display interfaces. The following screens are available:
- Substation overview screen
- Circuit screen for each feeder
- Alarm screen
- Sequence of events screen
- Real-time trend and historical data screens
- Protection options screens
- System configuration screen

The following user interface features are supported:
- Real-time plant monitoring
- Logical zooming on bay views
- Secured breaker control
- Real-time measurement

- Power and energy calculation
- Alarm management
- Sequence of time-tagged events
- Report handling and printing
- Data archiving and trending
- Secure operator access
- Reading and setting relay protection parameters
- Disturbance record extraction and analysis
- Supervision and diagnostics

System security is based on the use of multiple user names and passwords.

Alstom recommends the following minimum requirements for the PC:

- Pentium 300 MHz
- 6 GB hard disk
- 128 MB RAM
- 15″ to 21″ SVGA color monitor
- Keyboard and mouse
- Printer
- Serial communications ports and modem(s)

The MiCOM S10 software provides communication drivers and protocol converters for different relays. Communication protocols supported for the substation environment are MODBUS, Courier and IEC 60870-5-103.

Upper communication to a LAN/WAN is supported by MODBUS and IEC 60870-5-101.

Courier protocol has been developed by Alstom specifically for protection relays. Courier is based on a master/slave protocol whereby relays only respond when requested to do so by the master station. Devices are linked using K-Bus, which consists of a screened twisted pair of wires, an arrangement developed for communicating in electrically hostile environments. Courier offers a speed of 64 kbps.

Functions available using Courier include relay setting changes, polling measurement values, configuration, disturbance and event record extraction, and remote circuit-breaker operations. Access to the relays is provided from the master station.

Protection intertripping and blocking signals are routed via separate conventional wiring.

Critical functions, such as protection functions and time-tagging, are executed on the object level. This level ideally consists of IEDs with integrated protection and control functions. The use of the standard protocols MODBUS and IEC 60870-5-103 provides a flexible and open communications environment.

Capacity
- Up to 32000 data items per substation
- Up to 8 sub-networks per PC
- Up to 32 devices per sub-network

The MiCOM S10 is a typical type 1 system, as discussed in Chapter 10.

11.6.5 MiCOM S100

The 'MiCOM S100 Integrated Substation Protection and Control System' has been designed specifically for medium and large substations. The system is illustrated in Figure 11.11.

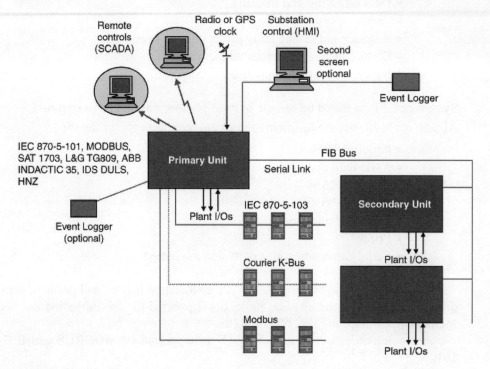

Figure 11.11
ALSTOM MiCOM S100 system

The heart of the MiCOM S100 system is the substation RTU. The RTU can handle both digital and analog inputs/outputs. Data capture and output controls are accomplished by I/O boards or via protective relays and IEDs. The number of objects and related functions controlled by the substation RTU are:

Digital points	4032
Measurements	480
Meterings	144
Output controls	1024
Analog setpoints	60
Digital setpoints	51
Tap changer positions	51
Bays	64

Up to a total of 5000 wired I/Os and up to 256 devices are supported. Resolution and accuracy for time-tagged events is 1 ms.

Up to 1000 automatic sequences, each consisting of up to 20 basic commands, which are triggered by an event, can be defined during configuration.

A FIFO buffer of 100 000 events is managed and stored on hard disk.

The following protocols are supported by SiCOM S100 for substation communications:
- ModBus
- K-Bus Courier
- IEC 60870-5-103
- IEC 60870-5-101

MiCOM S100 can communicate with up to three remote master stations, using separate communication protocols and databases. Channel redundancy can also be managed. Protocols supported for this function are:
- IEC 60870-5-101
- ModBus
- SAT1703
- Siemens/SAT
- Landis & Gyr TG809
- ABB P214
- HNZ 66S15
- HNZ ELENAS
- Other private protocols may be supported on request

MiCOM S100 is a typical type 4 system, as described in Chapter 10.

11.6.6 PSCN 3020

The PSCN 3020 Integrated Protection and Distributed Digital Control System provides combined protection, control and supervision for large transmission and distribution control systems. The system is organized around a double fiber optic local communication ring network and the use of bay modules (BM9100), which communicate with ALSTOM's range of protective relays and which also perform the automation functions.

Substation communications between the bay modules and protective devices are via serial link using MODBUS, K-Bus Courier, OPN-BUS, or IEC 60870-5-103. The BM9100 bay module provides hardwired interfaces to relays without communication abilities. The whole range of ALSTOM protective relays can be used in the PCSN 3020, as well as compatible relays from other manufacturers.

Communications with the remote control centres are via MODBUS, DNP3, IEC 60870-5-101, CONITEL, or CDC2.

The system consists of a double redundant ring principle. Up to 5 double rings can be accommodated with 240 devices per ring. Maximum distance between two connections is 2000 m, and the maximum circumference of one ring is 26 km.

The BM9100 bay module has been designed to control the electrical processes by the use of direct input/output modules, protective relays or IEDs. Data exchanged with the network include status, controls, measurements, time synchronization, parametering, disturbance records, etc. Automation functions are performed using information retrieved locally, through connected protection, or from other bay modules. It provides interlocking and sequences like feeder changeover or tap changer management.

All digital inputs and analog thresholds are time-stamped with a 1 ms precision. A local mimic is integrated within the BM9100 front panel for manual control. Hardware capacity of a single bay unit includes 96 digital inputs, 144 digital outputs, 24 analog inputs and up to 9 I/O boards.

The HMI provides full operator access, including remote relay settings, electrical supervision and control, real-time data display, sequence of event management, automatic disturbance record retrieval, data archiving and trends, alarm management, and system time synchronization. The HMI is PC architecture based on Windows NT with network communication facilities and modular software allowing easy addition of communication protocols. The HMI may be located locally in the substation or remotely in a nearby control station.

The redundant fiber ring network supports a network speed of 3.5 Mbps with a typical cycle duration of 2 ms for 40 devices. A deterministic token ring communication architecture, without master or slave, is used. High system reliability is obtained with no possibility of message collisions. The redundant network with a self-healing mechanism maintains communication even in case of a break in the fiber optic or device failure.

The PCSN 3020 system is a typical type 2 system, as described in Chapter 10.

11.6.7 SPACE 2000

The SPACE 2000 architecture corresponds to the PCSN 3020 system, but is designed around a flexible Ethernet TCP/IP communication network backbone, intended to be compatible with the future IEC 61850 standard.

This system is illustrated in Figure 11.12.

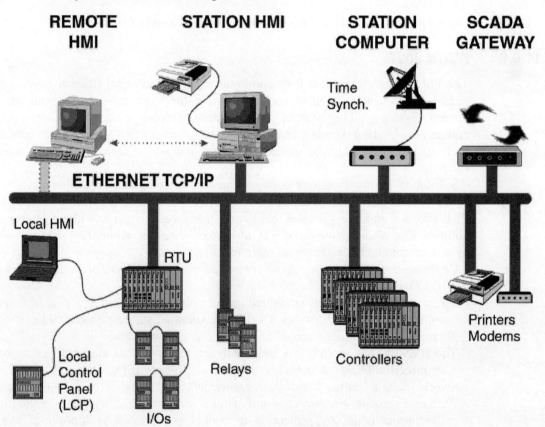

Figure 11.12
ALSTOM SPACE 2000 system

Multi-functional 'bay computers' perform integrated control, protection and condition monitoring functions for one or more bays. IEDs such as protective relays are connected to the bay computers by Legacy communication protocols such as IEC 60870-5-103, MODBUS or K-bus Courier.

Other 'IEDs' like transformer monitors, tariff meters, power quality monitors, printers, modems, etc are directly connected to the Ethernet TCP/IP backbone.

The system has been designed for future communication standards, while providing a migration path to cope with the progressive retrofit of existing substations.

12

Practical considerations

12.1 Justification

Equipment becoming obsolete and cost reductions are the major driving factors for modernizing electrical substations.

Equipment may become too expensive to maintain, obtaining spare parts may become a major headache, and excessive downtime may be caused by equipment failure. Changing conditions and/or operating strategies may necessitate the use of more modern equipment.

Self-checking and condition monitoring capabilities of modern IEDs are significantly reducing the need for periodic maintenance, and give early warning of potential failures.

Functional integration of IEDs allows one device to replace a collection of elementary devices, like various protection relays, disturbance recorders, metering devices, etc. This reduces substation maintenance costs and copes with the need for improved functionality.

Huge reductions in operating costs may be obtained with more efficient network operations and reduction in personnel costs, as extensive real-time information is available at the central control center.

12.2 Basic strategies

There are two basic strategies that can be followed when refurbishing a substation:
- The big bang approach, i.e. remove and rebuild everything. This can mean replacing existing switchgear as well, or replacing all the protection and control devices. This approach provides a homogeneous solution and is less expensive than the progressive change method, since there are no cohabitation interfaces to be developed and a single commissioning and test procedure are required. However, the big bang approach is capital intensive, and requires a whole substation to be taken out of commission.
- Progressive change method, which is subdivided into two directions:
  - Functions: Protection or control may be changed separately

– Bays: One bay or a set of bays may be renewed, keeping the other ones untouched

The advantages of this approach include:
- Adaptation to yearly budgets
- Continuous improvement reflecting both changing user needs and technology changes
- Rapid installation on each bay
- Progressive change is the most common scenario and requires a migration strategy, defining different steps to be followed to match the operational objectives.

12.3 Constraints

The main constraints when refurbishing a substation are:
- Minimization of the down time. Refurbishment of one bay may have an influence on the other bays.
- Cohabitation between various generations of devices, since investments have usually been progressive and should be optimized.
- Compatibility of operation with the previous generation of devices in order to be easily used by the operators. For example, the old mimic may have to be emulated, using the same security rules.

12.4 Competence management

Conventional systems usually had a clear separation between protection and control. Specialists for each discipline would make decisions regarding the choice of devices, manufacturers, installation and configuration.

Modern digital devices have integrated protection and control features. Added functionality also means more complex configuration and commissioning requirements. This places different demands on the competencies of technical specialists, and the question arises whether or not to keep and train specialists for the configuration, installation and maintenance of the new system within the utility/industry.

This will generally depend on the number of substations: When few substations are concerned it will generally be more economical to externalize the service; but when more substations are involved, or if one or more substations are refurbished regularly, it may be worthwhile to train specialists in the specific product range.

12.5 Electrical protection

When considering a *power system automation* system, the question inevitably arises: What to do with the existing protection?

The answer to this question is not so easy. It will depend a lot on the type of existing protection, as well as the type of power system automation system that will be installed, and the protection and control philosophy of the user.

The following are some guidelines that may be followed:

First of all, it should be kept in mind that the aim of a *power system automation* system is to have an integrated protection, control, metering and monitoring system. Therefore, the protection devices have to communicate to the rest of the system. If this is not the case, the system in question may be called a control and monitoring system, or something similar, but not, by definition, a power system automation system. In that case, protection will be independent from the rest of the system, and upgrading of the protection may be considered in isolation. This scenario will not be considered further, and the following discussion will focus on an integrated power system automation system, which is the topic of this text.

- Electromechanical relays
  These relays do not have any data communication capabilities, and should be replaced (concerning the switchgear bays that need to be part of the power system automation system).

- Static electronic relays
  These relays were analog relays, not microprocessor-based, and did not have data communication capabilities. There are not many of them still left in service, but if there are, they should be replaced.

- Digital relays
  First generation digital relays have limited data communication capabilities, generally only available via a front serial port, which was intended only to facilitate installation and commissioning. Strictly speaking, it should be possible to partly integrate certain types of these relays to a power system automation system. However, their functionality would be severely limited, restricted to their protection functions only, supporting very few of the functions the power system automation system would demand from them.

The cost and effort to integrate 1st generation digital relays to the system would be considerable, justifying their replacement in most cases.

Second generation relays supported SCADA communications. The question whether to replace them or not, will depend on the approach of the specific customer. Functionality of second generation relays is generally less than third generation relays. If increased protection functionality and/or reduced maintenance costs were the prime justification for implementing a power system automation system or modernizing a substation, these relays will probably be replaced. Sometimes they are kept as backup protection, but this defeats the maintenance objective, and the versatility of IEDs ensures that flexible backup protection schemes can be devised without additional (older) equipment.

If adding local and remote control functionality to the substation was the main justification for implementing a power system automation system, the existing protection may still be sufficient. In this case, it will depend on the type of system to be installed, i.e. if protection and control will be integrated in one device. If not, a bay controller may be installed, with the protection either communicating to the bay controller, communications processor, or RTU.

If the protection devices are not replaced during the first phase of a power system automation project, this may be considered at a following stage. The added functionality of modern devices may influence a decision to upgrade faster than originally planned.

12.6. Suppliers

The main advantages of installing products from a single supplier include a reduction of the customer's investment in training costs and optimized integration between devices.

Advantages of choosing multiple suppliers are better cost/function ratio and increased technical knowledge gained by the customer.

Compatibility, specifically of communication abilities, should be carefully considered when installing products from different suppliers. This should become less and less a concern with future communications standards from the IEC and IEEE providing a common platform for different manufacturers' products.

12.7 Power system automation and the Internet

Several utilities in the USA are turning their substations into secure websites. Personnel can have 'immediate' access to the substation from any location, using a PC or laptop and a web browser. US utilities are also deploying traditional SCADA functions over the Internet.

The substations are presented in a HTML index page. The user may then select the desired substation and, once he has passed the security controls, will have the following options:

- Monitoring the current real-time state in the substation
- Browsing the events and alarms that have been registered
- Retrieve stored data
- Executing commands on the substation devices

These developments are driven by the need to reduce maintenance costs in order to be more competitive in a deregulated supply market.

Cost advantages can certainly be obtained by following this approach. However, three main restrictions necessitate extreme caution when considering using the Internet for substation monitoring and control:

- Undoubtedly the most crucial consideration is security to prevent unauthorized access to the substation via the Internet,
- Data integrity is vital when commands are executed, and
- Response times, specifically regarding the execution of commands, are a severe limitation of the Internet at present.

These, and other, elements of the Internet are discussed in Appendix 1.

12.8 Summary

It is impossible to set '10 golden rules' for implementing a power system automation system. This will depend on the user's protection and control philosophy, existing equipment already installed, plant operating procedures, user preferences, etc.

A final word: The power system automation should be designed by, first of all, clearly defining the technical requirements for the system at the hand of the protection and control philosophy. The system should not be designed, which is often the case, around the limitations of equipment already chosen or installed.

Existing equipment will certainly influence the way a *power system automation* will be installed, demanding cohabitation with new devices and defining the migration path to be followed, but it should not influence the ultimate goal of the project. This should be defined by what the user wants out of the system at the end of the day.

Appendix A

The Internet

A1.1 History and background

The Internet is sometimes confused with the World Wide Web, also known as WWW, www or W3. Whereas the Internet provides the infrastructure that allows computers across the globe to interconnect, the web is software that 'lives' on the Internet, providing a graphical interface or 'doorway' to the Internet. The web server runs on a host computer, in a similar way as a mail or print server.

The Internet was originally known as the Advanced Research Projects Agency Network (ARPANET)) and was built by Bolt, Beranek, and Newman Inc. (BBN). This system operated from 1969 through to 1990 and was the template, or design base for TCP/IP (see Chapter 7, section 7.5).

However, by the late 1980s there was still no common user-friendly interface to the Internet. In 1989 Tim Berners-Lee, a scientist working at the European Organization for Nuclear Research (CERN) in Switzerland, conceived the idea of the WWW for the purpose of aiding research, collaboration and communication amongst colleagues within CERN. The result proved to be so popular that the web gained worldwide acceptance. The rest is history.

A web browser allows web pages (which are, in fact, files) resident on any web server to be selected and viewed as requested by a remote user. The original web browser was fairly unsophisticated and was driven by command line keyboard inputs. Subsequent mouse-based browsers were developed and graphics support was added. Information is accessed by pointing and clicking on hyperlinks-, images or words that enable access to new information.

There are two types of hyperlinks, namely hypertext and hypermedia.

Hypertext is the most commonly found hyperlink. Whether using a browser, such as Netscape Navigator and Microsoft Internet Explorer, or a word processor such as Corel WordPerfect 8 and Word 97 (and subsequent releases thereof) the hyperlinks can be shown in different colors and styles in order to make them more visible. Clicking on the hyperlink establishes a connection to the particular web page.

Hypermedia is another type of hyperlink technology used extensively today. Originally hypermedia meant that one could click on, say, a picture in order to access a particular web page. Nowadays it also means that different types of media (images, sound, animation) can actually be linked to information.

A1.2 Overview of HTML

All web pages are created using a special language known as Hypertext Markup Language (HTML), which allows one to organize text, graphics, animation and sound into documents that a browser can understand. HTML is the 'glue' that holds the web together; it is the language that makes hypertext and hypermedia possible.

Although HTML is indeed a language, it is not the type of programming language typically associated with computers and software development (such as Pascal or C++). Instead, HTML is a user-friendly markup language that practically everyone can begin using within a day or two.

Markup languages define a formal set of rules and procedures for preparing text to be electronically interpreted and presented. With HTML, one surrounds text and references to files with special directives known as tags. Tags are used to specify how the text or files are supposed to appear when viewed with a web browser; they are used to 'markup' the document in a way that the web browser understands how to deal with. Using tags to mark up a document for electronic publication is easy. One can take a standard word processor document, add some HTML, and thereby create a web page. The whole process can take less than 15 minutes when creating simple pages.

What really makes HTML powerful is its ability to organize any number of files onto a single page. Files appearing on a page may be physically located on the same computer as the page itself, or anywhere else on the web. Each file is stored independent of the pages in which they appear; that is, files are not stored inside of the web pages that display them. Instead, HTML merely references, or points to, these files, telling the browser exactly where they are located. A web page is nothing more than a text file that may contain references to any number of image, animation, and sound files that the browser will retrieve, assemble and display when that page is accessed.

A1.3 Overview of HTTP

HTTP (Hypertext Transfer Protocol) is the protocol that enables the connection between a web server and a client. By using a browser one could, for example, access IDC's website at www.idc-online.com by using the browser's 'go to' command and entering http://www.idc-online.com. Typing www.idc-online.com is usually sufficient since most browsers would by default use the http protocol to access the website.

The first web page displayed would be the home page or top level web page. From here on one would navigate to other associated pages by clicking on hyperlinks.

It is not imperative to use the http protocol in order to display the contents of a web page. One could simply dial up a Telnet connection to the web server, for example by invoking Telnet and connecting to www.idc-online.com at port 80. (Port 80 is used since web servers, by default, listen out for connection on port 80.) Alternatively, one could type >telnet www.idc-online.com 80 under the DOS command prompt. The only problem when not using http is that the page would not be interpreted and displayed as a typical web page as we know it, but as a listing of the html code only.

At its most basic level, the HTTP protocol consists of a single connection and a single command line delivered to a web server residing at a specific IP address. A problem with the real-life situation is that a single web server could hold several hundred websites, each one theoretically needing its own IP address. In addition to this, each website could have several dozens of web pages, each page requiring a separate connection with the client. To overcome this problem the HTTP 1.1 specification (and upwards) allows the administrator to assign a virtual host, which allows the website to appear to the outside world as a single entity with only one IP address.

A1.4 Java

Web pages made with HTML are static. Java programs, called applets, bring pages to life with animation, sound, and other forms of executable content. Unfortunately, Java applets are usually 'Plug and Play' since they cannot easily be modified. There are several reasons for this.

- The Java language is rather complicated and before one can write or modify an applet, the language first has to be mastered.
- It is not possible to view the source code of the applet.
- Applets cannot be (or rather should not be) downloaded without permission of the author.
- Even if the end user is capable of writing applets, an existing applet cannot be modified unless all parameters have been provided.

The major advantage of an applet is the fact that only the required part of an exe program can be downloaded to the user with the information required. Therefore, the user need not have the required software installed on his PC to view the information. Applets also make web pages interactive, and are used more and more for tasks like Internet banking.

A1.5 CGI

CGI (common gateway interface) looks like HTML and can accomplish some of the things that Java applets and JavaScript can, but it has distinct shortcomings. Java and JavaScript are therefore expected to make CGI obsolete.

- Compared to Java, CGI is difficult to learn.
- It is mainly used for entering alphanumeric text (e.g. parameters for search engines or credit card numbers for on-line purchases). However, it does not really interact with the user but rather submits the entered information to the web server for further processing.

A1.6 Scripting: JavaScript

Scripting languages have been around since the inception of programming languages and computers, and are commonly known as macros. Macros outline a list of predetermined steps that a spreadsheet performs when that macro is invoked – making macros little more than special purpose scripts. A macro in a spreadsheet is a therefore a form of scripting language.

JavaScript is a scripting language for the World Wide Web, developed by Netscape Communications Corp and Sun Microsystems, and is not to be confused with Java itself. Whereas Java is a full-blown programming language meant to be used by experienced software developers, JavaScript is a scripting language for the less experienced and consists of easy-to-understand English-type language. In terms of difficulty, scripting languages fall somewhere between markup languages, such as HTML, and full-blown programming languages, such as Java. Scripting languages provide much more than the ability to prepare documents for electronic publication, yet are not nearly as powerful as true programming languages. Scripting languages are, in essence, mini-programming languages for the average person.

Scripting languages fill a void left by programming languages. Whereas programming languages are used to create software products (such as word processors, spreadsheets, web browsers, applets), scripting language let the end user control such programs. In fact, a scripting language is defined as a relatively easy-to-use programming language that allows the end user to control existing programs. A software engineer creates a program using a programming language like Java, and the end user gets to control the program using a scripting language like JavaScript.

A1.7   XML

XML stands for eXtensible Markup Language, and is a data format for structured document interchange on the web. Like HTML, it is a markup language derived from SGML. It differs from HTML in that it is best suited for organizing data, whereas HTML which was created to allow cross-platform formatting of information for display. Stated in another way; while HTML specifies how a document should be displayed, it does not describe what kind of information the document contains, or how it is organized. XML allows document authors to organize information in a standard way. It is said that 'XML does for data what HTML does for display'.

The development of XML is a public project headed by the World Wide Web Consortium and is not owned by a specific company. The group is only open to members of W3C member companies, but their work can be followed by viewing the w3c website.

A1.8   Server side includes

Most HTML documents are static: that is, the server just sends the client the requested file with no changes. Unless, of course, the file contains Java or JavaScript applets. Sometimes, however, the user might want the server to modify the file every time it is accessed. This might be desirable in, for example, the following cases:
- Updating a counter each time a file is accessed, and forwarding this value with the file.
- Including additional text files in a document.
- Including the 'Date last modified' in a file, or the current date and time.
- Including the output of a CGI program.

This can be done using server side includes. The server processes the file (this is called parsing) and then sends the result to the client. The server needs to know that the file includes 'server side includes' to be parsed, and this can be done by using the extension .shtml instead of .html.

A1.9   Perl

Perl (Practical Extraction and Report Language) is a text processing programming language having sophisticated pattern matching capabilities and flexible syntax, and is used for applications such as input/output, file processing, file management, process management and system administration tasks.

A1.10 Internet access

A1.10.1 Connecting a single host to the Internet

Connection to the Internet backbone is supplied by 'primary' Internet service providers (ISP's) such as AOL (America On-line), CompuServe and Internet Africa. ISPs outside of the USA are connected to the US Internet backbone as well as to ISPs on other continents through high-speed undersea (fiber optic) and satellite connections with a bandwidth of several tens or even hundreds of megabits per second. These ISPs also own the servers needed for functions such as user authentication, mail (POP3 and SMTP) and domain name system (DNS) services. Users can subscribe to, and directly access these ISPs.

There is also a proliferation of 'secondary' ISPs differing from the others in that they do not own their own international access, but lease it from the primary ISPs such as those mentioned above. The 'secondary' ISPs are geographically dispersed and connect to the main ISPs via high speed public or private switched network links, for example X.25 and E1/T1.

The ISPs supply the points through which the Internet can be accessed (the so-called points of presence or PoP) either on a regional or national level, e.g. Ozemail (ozemail.com) in Australia or Internet Africa (iafrica.com) in South Africa, or on a global level e.g. IBM Global Network (ibm.net).

The ISP's equipment at the point of presence consists of:

- A router (or routers) which route traffic to other ISPs and to the Internet backbone.
- A point-to-point protocol (PPP) server to provide Internet connectivity with multiple Internet users (subscribers) across serial telephone lines. Some ISPs also offer SLIP (serial link interface protocol) but SLIP has largely been superseded by PPP.
- Analog (dial-up or leased-line) modems and ISDN connections are required for user access as required. The modems are connected to the local POTS exchange through dedicated telephone lines, one per modem, with a so-called 'hunting line' at the exchange so that all modems can be accessed via the same telephone number.

Until recently these routers, modems and PPP servers were installed as discrete units. The current trend is to purchase them as integrated access servers, with the routing, dial-up server and modem functions in one box. The typical number of modems per access server is around 30 but this number can vary, and the number of ports can simply be increased by stacking additional units.

Users can access the ISP through several means. In all cases the user pays the ISP for the Internet access, as well as the telephone supplier for the connection to the ISP. Usually the connection can be accomplished as a 'local' call. Access methods include:

- Dial-up modem over a normal telephone connection. This is by far the most cost effective method for a single user or a small group of users but a serious drawback is lack of speed, not so much due to the bandwidth limitation of the user's telephone line or modem, but by the total demand imposed on the access server by all the users and the capacity of the link between the secondary and primary ISPs. Experienced 'web surfers' know that the best time to access the Internet is during the early hours of the morning when most

other users are asleep! Even a 56 kbps modem can often not accomplish a connection at higher than 24 kbps and even then the user can be fortunate to achieve a data download rate of more than a few kbps during peak hours.
- ISDN connection. This is also a dial-up service, but the communication is digital and the bandwidth between subscriber and ISP is substantially higher. The typical '2B + D' connection offers a 128 kbps bandwidth, and additional channels can be dialed-up if more bandwidth is required. Because of the higher performance, the charges for this service are substantially higher.
- Leased lines. These provide permanent connection to the ISP and are divided into two categories: analog and digital. Analog leased line modems use the same technology and therefore have the same speed limitations. At present analog leased line modems operate at typically 33.6 kbps to 56 kbps. Distance and noise are limiting factors, and analog leased lines are often only half-duplex, which means that traffic can only travel in one direction at a time. Digital leased lines (e.g. X.25) are faster, more reliable, and not limited by distance.
- Cellular (mobile) phone. Laptop computers can link up with a suitably equipped ISP without using a traditional telephone-type connection. Apart from the cellular phone rates usually being higher than normal dial-up rates, this connectivity solution may necessitate the purchase of a dedicated PCMCIA (also known as CardBus or PC-Card) interface in order to connect to the laptop, or a new infrared compatible cellular phone. Older cellular phones such as the Nokia 2110 have an external communications connector but need a special PCMCIA interface for a laptop. Newer models such as the Ericsson SH 888 and Nokia 6110 come equipped with a built-in PCMCIA interface and can communicate with the laptop either via infrared link or RS-232.

A1.10.2 Connecting remote hosts to corporate LAN

Larger organizations often have an existing in-house LAN with permanent access to the Internet. Over and above the need for Internet access, users may still rather want to log in to the corporate network as opposed to an ISP for the following reasons:
- They may wish to access corporate databases and file servers from home or whilst on the road.
- Remote customer and vendor access to restricted corporate information such as order status or purchasing data.
- Remote diagnostic and maintenance activities by system administrators.

The solution is the installation of a communication server (also called a PPP server) supporting at least the IP (preferably also IPX, for Novell Netware users) protocol families. This enables workstations to dial in over standard telephone lines using modems. The communication server answers the phone, authenticates the user and attaches the remote workstation to the LAN. Subject to security constraints, the remote user can then access all IP (and IPX) LAN-based resources including databases, file servers, web servers, and routers. Depending on the specific model, a communication-server typically supports between 1 and 32 hosts.

A1.10.3 Connecting multiple hosts to the Internet

A1.10.3.1 Connection via proxy server

This approach is ideal for a LAN with only a few hosts on it, for example a small office LAN or 2–3 networked PCs at home, which all need access to the Internet at the same time.

In general, a 'proxy' stands-in for something, or somebody. A paid-up member of an organization, unable to attend the AGM, could hand a proxy to another member to vote on her behalf. In the case of a network the proxy server is the machine with the connection to the Internet (e.g. via dial-up modem). The server runs special proxy software such as WinGate or WinProxy, which allows any other client computer on the network to forward its request, for something like a web page, to be handled on its behalf by the proxy server. The proxy server, in turn, downloads the web page and passes it back to the client in a manner that is transparent to the user.

Proxy servers can usually handle only one protocol and are generally aimed at occasional dial-up Internet connection for small organizations. They are not intended for organizations where they would be key connection to the Internet.

No special configuration for the client machines are normally necessary, apart from informing Internet Explorer during setup that there is indeed a proxy server, what its IP address is, and at what port number it runs. Information regarding the latter will be obtained from the proxy server's documentation.

A1.10.3.2 Connection via NAT server (IP masquerading)

NAT, or network address translation (also referred to as IP masquerading) is intended for a permanent, 'heavy duty' connection to the Internet. Whereas this solution physically looks the same as proxy serving, it operates on a very different principle.

Its operation is entirely transparent to the rest of the network. Client computers on the network can use virtually any protocol; there is no special software and very little configuration required for them, apart from the normal TCP/IP setup. The only problem is that from the Internet point of view, there will be only one IP address and hence only one host visible on the network, namely the machine configured as the NAT server.

The client machines are configured to view the NAT machine as the default gateway (router), which is indeed what it is. The NAT server receives a packet from a client, replaces the IP address in the frame with its own, and forwards it onto the Internet. When a return message reaches the NAT gateway, it replaces the destination address with that of the client computer and forwards it on to its own subnet. Besides just translating addresses, NAT must also translate header information and packet checksums.

A1.10.3.3 Connection via IP sharer

An Internet IP sharer such as Micronet's SP86X is a hardware device that comes pre-programmed with a set of valid IP addresses. It acts as a DHCP server, automatically allocating IP addresses to each active station on the LAN.

It provides a firewall function and will automatically dial-up and disconnect depending on usage. Connection with the ISP is achieved via 56 kbps dial-up modems or 128 kbps ISDN. Depending on the model being used, 1, 2 or 4 modems can be connected in parallel, individual modems being activated or deactivated according to bandwidth requirement.

A1.10.3.4 Connection via UNIX or NT gateway

This is one of the easiest solutions for a large company wishing to give Internet access to all its members. A UNIX or NT host is setup as a gateway to the Internet. This solution requires at least a set of 254 Class C IP addresses, either permanently allocated to hosts, or dynamically allocated via a DHCP server which could run on the same machine.

The UNIX/NT machine is setup with two network adapters, i.e. as a 'multi-homed' host. The first adapter is connected to the internal LAN, the second to the Internet. This implies that the second adapter should have the necessary connectivity e.g. X.25 built-in. Each card will need its own permanent IP address, and each card will be configured in such a way that the 'other' card's IP address will be given as its default gateway. In this way, each card will pass on a received message to the other card.

A1.10.3.5 Connection via dedicated router

The simplest way of connecting a network to the Internet is via a dedicated 2-port IP router (also referred to as an Internet router). One port of the router will be, for example, an Ethernet port, to be connected to the local area network. The other port could be an X.25 WAN port, which will be connected to a public packet switching network. The X.25 link provides the connection to the ISP.

As in the previous case, a set of IP addresses is required, and these are allocated either permanently or via a DHCP server.

A1.11 Internet communications

A1.11.1 Introduction

A very interesting and significant global development is the changeover from traditional MAN/WAN architectures, used for linking company resources over large geographical areas, to Internet communication because of (a) the much lower cost involved and (b) simplicity of interconnection imposed by the necessity of standardizing on TCP/IP. This tendency is not only manifesting itself in the so-called information technology (IT) environment, but also in commerce and industry, particularly in the manufacturing and process control environments. It is therefore only logical that the Internet will also be used for telecommunications (voice, fax, video etc) on an ever-increasing scale.

A1.11.2 Hardware and software issues

The advantage of the current generation of Internet communications products lies therein that they co-exist on the already established Internet, PSTN (public switched telephone network) and PBX (private branch exchange) infrastructure. Internet communications products are predominantly software-based and in many cases, they are available either as freeware or shareware.

For the top-end products it may be necessary to purchase dedicated Internet interfaces for telephones or fax machines, but in these cases the end-users are typically medium to large enterprises and the capital outlay can be justified in terms of cost savings.

A1.11.3 Speed/bandwidth issues

As far as long-distance communication over the Internet is concerned, transmission speed can be a problem. For a modem-connected dial-up user the fastest modem in the world will not improve things much since the bottleneck is imposed by the bandwidth made available to the Internet service provider (ISP) and the number of simultaneous users competing for access to the ISP. In some cases the bandwidth of this data 'pipe' feeding the ISP is as little as 64 kbps, with 100 users connected at a given time: this translates to only 640 bps per user!

When it comes to increasing the available bandwidth within a LAN, however, there are several possibilities open to the LAN owner. Increasing the data transmission speed of the LAN (say, by upgrading from 10BaseT to 100BaseT) is one option, but not the only. Additional (and in some cases less costly) options are:

- Careful segmentation of a large flat network, with bridges, switches and routers.
- This reduces traffic interference as well as collisions.
- Cutting down on unnecessary broadcast packets (there are ways to accomplish this).

Minimizing the number of routers between a given workstation and the point where the LAN attaches to the WAN leads to fewer 'hops' across routers and thus reduces latency (time delays), which adversely affects voice and video transmissions.

Tasks that result in heavy network traffic, such as backups and large file transfers can be scheduled for off-peak periods in order to minimize interference with voice/video/fax traffic, which normally takes place during working hours.

A1.12 The security problem

A1.12.1 Overview

Although people tend to refer to the 'Internet' as one global entity, there are in fact three clearly defined subsets of this global network'; four, in fact, if you wish to include the so-called 'community network'. It just depends on where one conceptually draws the boundaries.

- In the center is the in-house corporate 'intranet', primarily for the benefit of the people *within* the organization.
- The intranet is surrounded by the 'extranet', exterior to the organization yet restricted to access by business partners, customers, and preferred suppliers.
- Third, and this is optional, there can be a 'community' layer around the extranet. This space is shared with a particular community of interest, e.g. industry associations.
- Finally, these three layers are surrounded by the global Internet as we know it, which is shared by prospective clients/customers and the rest of the world.

This expansion of the Internet into organizations, in fact right down to the factory shop floor, has opened the door to incredible opportunities. Unfortunately it has also opened the door to all sorts of pirates and hackers. Therefore, as the use of the Internet, intranets and extranets has grown, so has the need for security. The TCP/IP protocols and network technologies are inherently designed to be open in order to allow interoperability. Therefore, unless proper precautions are taken, data can readily be intercepted and altered – often without either the sending or the receiving party being aware of the security breach. Because dedicated links between the parties in a communication are often not established in advance, it is easy for one party to impersonate another party.

There is a misconception that attacks on a network will always take place from the outside. This is as true of networks as it is true of governments. In recent times the growth in network size and complexity has increased the potential points of attack both from outside and from within. Without going into too much detail, the following list attempts to give an idea of the magnitude of the threat experienced by intranets and extranets:

- Unauthorized access by contractors or visitors to a company's computer system.
- Authorized users (employees or suppliers) access unauthorized databases. For example, an engineer might break into the human resources database to obtain confidential salary information.
- Confidential information might be intercepted as it is being sent to an authorized user. A hacker might attach a network sniffing device (probe) to the network, or use sniffing software on his computer. While sniffers are normally used for network diagnostics, they can also be used to intercept data coming over the network medium.
- Users may share documents between geographically separated offices over the Internet or extranet, or 'telecommuters' users accessing the corporate intranet from their home computer can expose sensitive data as it is sent over the medium.
- Electronic mail can be intercepted in transmit, or hackers can break into the mail server.

Here follows a list of some additional threats:
- SYN flood attacks
- Fat ping attacks (ping of death)
- IP spoofing
- Malformed packet attacks (TCP and UDP)
- ACK storms
- Forged source address packets
- Packet fragmentation attacks
- Session hijacking
- Log overflow attacks
- SNMP attacks
- Log manipulation
- ICMP broadcast flooding
- Source routed packets
- Land attack
- ARP attacks
- Ghost routing attacks
- Sequence number prediction
- FTP bounce or port call attack
- Buffer overflows
- ICMP protocol tunneling
- VPN key generation attacks
- Authentication race attacks

These are not merely theoretical concerns. While computer hackers breaking into corporate computer systems over the Internet have received a great deal of press in recent years, in reality, insiders such as employees, former employees, contractors working onsite, and other suppliers are far more likely to attack their own company's computer systems over an intranet. In a 1998 survey of 520 security practitioners in US corporations and other institutions conducted by the Computer Security Institute (CSI) with the participation of the FBI, 44 per cent reported unauthorized access by employees compared with 24 per cent reporting system penetration from the outside!

Such insider security breaches are likely to result in greater losses than attacks from the outside. Of the organizations that were able to quantify their losses, the CSI survey found that the most serious financial losses occurred through unauthorized access by insiders, with 18 companies reporting total losses of $51 million as compared with $86 million for the remaining 223 companies.

The following list gives the average losses from various types of attacks as per the CSI/FBI 1998 Survey of Computer Security:

Type of Attack	Average Loss (millions of $)
Unauthorized insider access	$ 1.363
Theft of proprietary information	$ 1.307
Financial fraud	$ 656
Telecommunications fraud	$ 595
Sabotage of data or networks	$ 164
Spoofing	$ 128
System penetration by outsiders	$ 111
Telecom eavesdropping	$ 97
Denial of service	$ 77
Virus	$ 66
Active wiretapping	$ 49
Insider abuse of net access	$ 39
Laptop theft	$ 35
AVERAGE LOSS	**$ 215**

Fortunately technology has kept up with the problem, and the rest of this chapter will deal with possible solutions to the threat. Keep in mind that securing a network is a continuous process, not a one-time prescription drug that can be bought over the counter.

Also remember that the most sensible approach is a defense-in-depth ('belt-and-braces') approach as used by the nuclear industry. What is meant by this is that one should not rely on a single approach, but rather a combination of measures with varying levels of complexity and cost.

A1.12.2 Controlling access to the network

There are several ways of controlling access to a network. These include:
- Passwords
- Routers
- Firewalls
- Authentication

A1.12.2.1 Passwords
A company whose LAN (or intranet) is not routed to the Internet mainly has to face internal threats to its network. A password function is built into all networking software, but its function is mainly to 'keep the good guys out'.

A1.12.2.2 Routers
A router can be used as a simple firewall that connects the intranet to the 'outside world'. Despite the fact that its primary purpose is to route packets, it can also be used to protect the intranet.

In comparison to firewalls, routers are extremely simple devices and are clearly not as effective as firewalls in properly securing a network perimeter access point. However, despite their lack of sophistication, there is much that can be done with routers to improve security on a network. In many cases these changes involve little administrative overhead. There are two broad objectives in securing a router, namely:

- Protecting the router itself, and
- Using the router to protect the rest of the network.

Protecting the routers :

The following approaches can be taken.

- Keep the router software current. This could be a formidable task, especially for managers maintaining a large routed network who are likely to be faced with the prospect of updating code on hundreds of devices. It is, however, an essential one since operating routers on current code is a substantial step toward protecting them from attack and properly maintaining security on a network. In addition, new updated software revisions often provide performance benefits, offering more leeway to address security concerns without bringing network traffic to a halt.
- It is imperative for network managers to keep current on release notes and vendor bulletins. Release notes are a good source of information and enable network managers to determine whether or not a fix is applicable to their organization. In the case of a detected vulnerability in the software for a particular router, CERT advisories and vendor bulletins often provide workarounds to minimize risk until a solution to the problem has been found.
- Verify that the network manager's password is strong and make sure the password is changed periodically and distributed as safely and minimally as possible. More important, verify that all non-supervisory level accounts are password protected, to prevent unauthorized users from reading the router's configuration information.
- Allow Telnet access to the router only from specific IP addresses.
- Authenticate any routing protocol possible.
- From a security perspective, SNMP is a poor protocol to use. However, it does aid in managing the network. Defining a limited set of authorized management stations is always prudent. While this will not prevent spoofed address attacks at the router, it is nonetheless a simple step that should not be overlooked.

Protecting the network

- Logging. Logging the actions of the router can assist in completing the overall picture of the condition of the network. The ideal solution is to keep one copy of the log on the router, as well as one on a remote logging facility, such as *syslog*, since an attacker could potentially fill the router's limited internal log storage to erase details of the attack. With only remote storage, though, the attacker need only disrupt the logging service to prevent events from being recorded.
- Access control lists (ACLs). ACLs allow the router to reject or pass packets based on TCP port number, IP source address or IP destination address. Traffic control can be accomplished on the basis of (a) implicit permission, which means only traffic not specifically prohibited will be passed through, or

(b) implicit denial which means that all traffic not specifically allowed will be denied.

A1.12.2.3 Firewalls

Routers can be used to block unwanted traffic and therefore act as a first line of defense against unwanted network traffic, thereby performing basic firewall functions. It must, however, be kept in mind that they were developed for a different purpose, namely routing, and that their ability to assist in protecting the network is just an additional advantage. Routers, however sophisticated, generally do not make particularly intricate decisions about the content or source of a data packet. For this reason network managers have to revert to dedicated firewalls.

Firewalls are designed to sit on the boundary between an intranet and the rest of the world, monitoring both incoming and outgoing traffic, allowing only specific incoming and outgoing packets to pass and rejecting all other packets. This is not such an impossible task, since all TCP/IP communications is based on a port number that is contained in the TCP header. On the basis of the port number, a firewall can be instructed about who can transmit data, to what port they can transmit, and what sort of incoming connections are allowed on the network. One firewall is usually sufficient, but since a firewall only guards against attacks 'from the other side' and not from within, several of them might have to be deployed internally within an intranet if information on a particular part or 'region' has to be secured against other parts of the organization.

Firewalls are implemented in two ways – hardware-based and software-based. Hardware-based firewalls are dedicated self-contained 'firewalls in boxes' such as the Cisco PIX firewall, and are generally faster albeit more expensive. On the other hand, software firewalls are built by running firewall software such as WinGate, running on an existing computer under NT or UNIX. This solution is generally less costly, but slower.

Apart from the way they are implemented (i.e. hardware or software), firewalls can also be divided into two distinct types. The two most common types are packet filtering firewalls (also referred to as network layer firewalls) and proxy servers.

Network layer firewalls

Network layer firewalls deal mostly with routing rules. In other words, when a packet of data arrives at the firewall, it checks to see where the packet came from, where it is going, what it is used for, and then decides whether or not it is authorized. It monitors the actual content of data streams and the services exchanging these streams, while also checking for IP or DNS (domain name service) spoofing. The most distinguishing feature of a network layer firewall is its ability to allow IP traffic to pass through it. Network layer firewalls are almost completely transparent and anyone using the intranet will, generally, not even be aware of its presence. Unfortunately, this means that the intranet is probably going to need an assigned IP address block, which can be difficult to obtain.

These routers employ several advanced techniques, including dynamic IP address allocation, sequence number scrambling, DMZ (de-militarized zoning) and 'strikeback'. These techniques will now be discussed briefly in order to facilitate a better understanding of how these devices operate.

Dynamic IP address allocation

This is also known as natural address translation or NAT.

With NAT, the private IP addresses of machines inside the network are hidden from the outside world. They therefore need not be registered, and can be assigned by the system administrator. The firewall, on the other hand, has a built-in set of legitimate IP addresses, which are typically contained within one class C address.

An outward-bound packet sent by a host inside the intranet follows a default route to the inside interface of the firewall. Upon receipt of the outbound packet, the firewall extracts the host's source addresses (MAC and IP) and replaces it with its own MAC address and a globally unique IP number from the firewall's pool of available IP addresses. The packet therefore seems to originate from the firewall. Since the differences between the original and translated versions of the packet are known, the checksums are updated with a simple adjustment rather than complete recalculation, which saves time.

Since it seems, to the outside world, as if the message has originated from the firewall, any returned messages would be routed back to the firewall. The firewall inspects returning packets, and once it is satisfied with their legitimacy, it strips the allocated IP address, returns it to the available pool of IP addresses, and restores the IP and MAC addresses of the original sender before sending it off to the originating host.

After a user-configurable timeout period during which there have been no returned packets for a particular address-mapping, the firewall removes the entry, freeing the global address for use by another inside host. This is done so that a particular IP address will not be tied up indefinitely in the case of a packet getting lost along the way.

TCP sequence number randomization

Dynamic IP address allocation, while secure, is not port-specific and relies on a simple configuration table to track removed addresses. As a result, it does not provide absolute security because a spoofer could, theoretically, initiate a packet from outside the network that travels with a signal coming back through the configuration table; thus obtaining all addresses.

To remove this potential weakness of dynamic IP address allocation, firewalls can capture the TCP sequence numbers and port numbers of originating TCP/IP connections. In order for spoofers to penetrate the firewall to reach an end server, they would need not only the IP address, but the port number and TCP sequence numbers as well.

To minimize the possibility of unauthorized network penetration, some firewalls also support sequence number randomization, a process that prevents potential IP address spoofing attacks, as described in a security advisory (CA-95:01) from the Computer Emergency Response Team (CERT). Essentially, this advisory proposes to randomize TCP sequence numbers in order to prevent spoofers from deciphering these numbers and then hijacking sessions. By using a randomizing algorithm to generate TCP sequence numbers, the firewall then makes this spoofing process extremely difficult, if not impossible. In fact, the only accesses that can occur through this type of firewall are those made from designated servers, which network administrators configure with a dedicated 'conduit' through the firewall to a specific server – and that server alone.

DMZ's (de-militarized zones)

Most firewalls have two ports, one connected to the intranet and the other to the outside world. The problem arises: on which side does one place a particular (e.g. WWW, FTP or any other application) server? On either side of the firewall the server is exposed to attacks, either from insiders or from outsiders. Some firewalls have a third port, protected from both the other ports, leading to a so-called DMZ or de-militarized zone. A server attached to this port is protected from attacks, both from inside and outside.

Strike back intruder response

Some firewalls have a so-called intruder response function. If an attack is detected or an alarm is triggered, it collects data on the attackers, their source, and the route they are using to attack the system. They can also be programmed to automatically print these results, e-mail them to the designated person, or initiate a real-time response via SNAP or a pager.

Application layer firewalls (proxy servers)

Application layer firewalls (proxy servers) perform basically the same function as network layer firewalls, although in a slightly different way. Basically, an application layer firewall acts as an ambassador for a LAN or intranet, the Internet. Proxies tend to perform elaborate logging and auditing of all the network traffic intended to pass between the LAN and the outside world, and can cache (store) information such as web pages so that the can client accesses it internally rather than directly from the web.

A proxy server or application layer firewall will be the only Internet connected machine on the LAN. The rest of the machines on the LAN have to connect to the Internet via the proxy server, and for them Internet connectivity is just simulated.

Because no other machines on the network are connected to the Internet, a valid IP address is not needed for every machine. Application layer firewalls are very effective for small office environments that are connected with a leased line and do not have allocated IP address blocks. They can even perform a dial-up connection on behalf of a LAN, and manage e-mail and any other Internet requests.

They do, however, have some drawbacks. Since all hosts on the network have to access the outside world via the proxy, any machine on the network that requires Internet access usually needs to be configured for the proxy. A proxy server hardly ever functions at a level completely transparent to the users. Furthermore, a proxy has to provide all the services that a user on the LAN uses, which means that there is a lot of server type software running for each request. This results in a slower performance than that of a network layer firewall.

A1.12.3 Intrusion detection systems (IDS)

Intrusion detection is a new technology, which enables network and security administrators to detect patterns of misuse within the content of their network traffic. IDS is a growing field and there are several excellent intrusion detection systems available today, not just traffic monitoring devices. Network General (the inventor of the Sniffer, now known as Networks Associates after its recent merger with MacAfee) sells a product called CyberCop. StorageTek sells a combination packet filter and intrusion detection system under the name of NetSentry (this combines their BorderGuard security device with the NetRanger IDS). The WheelGroup NetRanger itself is capable of automatically setting filters on the BorderGuard devices as well as Cisco routers, based on where (in the enterprise network) they are located and a company's real-time policy enforcement needs. NetSolve (ProWatch Secure) and IBM (Emergency Response Service) both offer intrusion detection monitoring and response services as well.

These systems (and services) are capable of centralized configuration management, alarm reporting, and attack info logging from many remote IDS sensors. ID systems are intended to be used in conjunction with firewalls and other filtering devices, not as the only defense against attacks.

There are two ways that intrusion detection is implemented in the industry today: host-based systems and network-based systems.

A1.12.3.1 Host-based IDS

Host-based intrusion detection systems use information from the operating system audit records to watch all operations occurring on the host on which the intrusion detection software has been installed. These operations are then compared with a pre-defined security policy. This analysis of the audit trail imposes potentially significant overhead requirements on the system because of the increased amount of processing power required by the intrusion detection software. Depending on the size of the audit trail and the processing power of the system, the review of audit data could result in the loss of a real-time analysis capability.

A1.12.3.2 Network-based IDS

Network-based intrusion detection, on the other hand, is performed by dedicated devices (as opposed to software running on hosts) that are attached to the network at several points and passively monitor network activity for indications of attacks. Network monitoring offers several advantages over traditional host-based intrusion detection systems. Because intrusions might occur at many possible points over a network, this technique is an excellent method of detecting attacks which may be missed by host-based intrusion detection mechanisms.

The greatest advantage of network monitoring mechanisms is their independence from reliance on audit data (logs). Because these methods do not require input from any operating system's audit trail they can use standard network protocols to monitor heterogeneous sets of operating systems and hosts.

Independence from audit trails also frees network monitoring systems from possessing an inherent weakness caused by the vulnerability of the audit trail to attack. Intruder actions which interfere with audit functions or which modify audit data can lead to the prevention of intrusion detection or the inability to identify the nature of an attack. Network monitors are able to avoid attracting the attention of intruders by passively observing network activity and reporting unusual occurrences.

Another significant advantage of detecting intrusions without relying on audit data is the improvement of system performance, which results from the removal of the overhead imposed by the analysis of audit trails, a process that degrades the performance of the system. In addition, techniques that move the audit data across network connections reduce the bandwidth available to other functions. Network monitoring techniques can increase performance of networks by 5 to 20 per cent compared to audit-based systems.

A1.12.3.3 Network flight recorder

An interesting variation on the theme is NFR (network flight recorder) which does not necessarily look for intrusions but provides information about the growth of the work, its usage patterns and potential misconfigurations. It can retain copies of all clients/partner communications, statistics about web surfing activity through the firewall, or records about who logged into the main frame, when and how long. A typical NFR system runs on a PC with a hard disk sized on how much data the system administrator wants to retain. Accessing the NFR's data-store is performed using a web browser that supports Java and the secure sockets layer. It is completely end-user programmable but includes a number of standard recording packages that gather basic statistics, watch firewalls, and track user activity.

A1.12.4 Security management

A1.12.4.1 Certification

Certification is the process of proving that the performance of a particular piece of equipment conforms to the laid-down policies and specifications. Whereas this is easy in the case of electrical wiring and wall sockets, where Underwriters' Laboratory can certify the product, it is a different case with networks where no official bodies and/or guidelines exist.

If one needs a certified network security solution, there are only two options viz.

- Trusting someone else's assumptions about one's network
- Certifying it oneself

It is possible to certify a network by oneself. This exercise will demand some time but will leave the certifier with a deeper knowledge of how the system operates.

The following are needed for self-certification:

- A management directive which favors security
- A security policy (see next section)
- Some basic knowledge of TCP/IP networking
- Access to the web
- Time

To simplify this discussion, we will assume we are certifying a firewall configuration. Let us look at each individually.

- Management directive which favors security

  One of the biggest weaknesses in security practice today is the large number of cases in which a formal vulnerability analysis finds a hole that simply cannot be fixed. Often the causes are a combination of existing network conditions, office politics, budgetary constraints, or lack of management support. Regardless of who is doing the analysis, management needs to clear up the political or budgetary obstacles that might prevent implementation of security.

- Security policy

  In this case, 'policy' means the access control rules that the network security product is intended to enforce. In the case of the firewall, the policy should list:

- The core services that are being permitted back and forth
- The systems to which those services are permitted
- The necessary controls on the service, either technical or behavioral
- The security impact of the service
- Assumptions that the service places on destination systems
- For example, web traffic permitted in through a firewall to an internal web server implicitly assumes that the web server's CGI scripts will be secure.
- Basic TCP/IP knowledge

  Many firewalls expose details of TCP/IP application behavior to the end user. Unfortunately, there have been cases where individuals bought firewalls and took advantage of the firewalls' easy 'point and click' interface, believing they were safe because they had a firewall. One needs to understand how each service to be allowed in and out operates, in order to make an informed decision about whether or not to permit it.

- Access to the web

  When starting to certify components of a system, one will need to research existing holes in the version of the components to be deployed. The web, and its search engines, are an invaluable tool for finding vendor-provided information about vulnerabilities, hacker-provided information about vulnerabilities, and wild rumors that are totally inaccurate. Once the certification process has been deployed, researching the components will be a periodic maintenance effort.

- Time

  Research takes time, and management needs to support this and to invest the time necessary to do the job right. Depending on the size/complexity of the security system in question, one could be looking at anything between a day's work and several weeks.

A1.12.4.2 Information security policies

The ultimate reason for having security policies is to save money.

This is accomplished by:
- minimizing cost of security incidents; accelerating development of new application systems
- justifying additional amounts for information security budgets
- establishing definitive reference points for audits

In the process of developing a corporate security consciousness, one will, amongst other things, have to:
- educate and train staff to become more security conscious
- generate credibility and visibility of the information security effort by visibly driving the process from a top management level
- assure consistent product selection and implementation
- coordinate the activities of internal decentralized groups

The corporate security policies are not only limited to minimize the possibility on internal and external intrusions, but also to :
- maintain trade secret protection for information assets
- arrange contractual obligations needed for legal action
- establish a basis for disciplinary actions
- demonstrate quality control processes for example ISO 9000 compliance

The topics covered in the security policy document should, for example, include:
- web pages
- firewalls
- electronic commerce
- computer viruses
- contingency planning
- Internet usage
- computer emergency response teams
- local area networks
- electronic mail
- telecommuting
- portable computers
- privacy issues
- outsourcing security functions
- employee surveillance
- digital signatures
- encryption
- logging controls
- intranets
- microcomputers
- password selection
- data classification
- telephone systems
- user training

In the process of implementing security policies, one need not re-invent the wheel. Products such as *Information Security Policies Made Easy* are available in a hardcopy book and CD-ROM. By using a word processing package, one can generate or update a professional policy statement in a couple of days.

A1.12.4.3 Security advisory services

There are several security advisory services available to the systems administrator. This section will deal with only three of them, as examples.

- Microsoft

  All software vendors issue security advisories from time to time, warning users about possible vulnerabilities in their software. A particular case in point is Microsoft's advisory regarding the Word97 template security, which was issued on 19 January 1999. This weakness was exploited by a devious party who subsequently devised the Melissa virus. See section 14.6 for a web address.

- CERT

  The CERT co-ordination center is based at the Carnegie Mellon Software Engineering Institute and offers a security advisory service on the Internet. Their services include:
  - CERT advisories
  - Incident notes
  - Vulnerability notes
  - Security improvement modules

  The latter includes topics such as:
  - detecting signs of intrusions
  - security for public websites
  - security for information technology service contracts
  - securing desktop stations
  - preparing to detect signs of intrusion
  - responding to intrusions
  - securing network services

These modules can be downloaded from the Internet in PDF or PostScript versions and are written for system and network administrators within an organization. These are the people whose day-to-day activities include installation, configuration, and maintenance of the computers and networks.

Once again, a particular case in point is the CERT/CC CA-99-04-MELISSA-MICRO-VIRUS.HTML dated March 27, 1999 which deals with the Melissa virus which was first reported at approximately 2:00 pm GMT-5 on Friday, 26 March 1999. This example indicates the swiftness with which organizations such as CERT react to threats.

- CSI
  CSI (The Computer Security Institute) is a membership organization specifically dedicated to serving and training the information computer and network security professionals. CSI sponsors two conferences and exhibitions each year: NetSec in June and the CSI Annual in November. CSI also hosts seminars on encryption, intrusion, management, firewalls and awareness. They also publish surveys and reports on topics such as computer crime and information security program assessment.

A1.12.5 The public key infrastructure (PKI)

A1.12.5.1 Introduction to cryptography

The concept of securing messages through cryptography has a long history. Indeed, Julius Caesar is credited with creating one of the earliest cryptographic systems to send military messages to his generals.

Throughout history, however, there has been one central problem limiting widespread use of cryptography. That problem is key management. In cryptographic systems, the term key refers to a numerical value used by an algorithm to alter information, making that information secure and visible only to individuals who have the corresponding key to recover the information. Consequently, the term key management refers to the secure administration of keys to provide them to users where and when they are required.

Historically, encryption systems used what is known as symmetric cryptography. Symmetric cryptography uses the same key for both encryption and decryption. Using symmetric cryptography, it is safe to send encrypted messages without fear of interception, because an interceptor is unlikely to be able to decipher the message. However, there always remains the difficult problem of how to securely transfer the key to the recipients of a message so that they can decrypt the message.

A major advance in cryptography occurred with the invention of public-key cryptography. The primary feature of public-key cryptography is that it removes the need to use the same key for encryption and decryption. With public-key cryptography, keys come in pairs of matched 'public' and 'private' keys. The public portion of the key pair can be distributed in a public manner without compromising the private portion, which must be kept secret by its owner. Encryption done with the public key can only be undone with the corresponding private key.

Before the invention of public-key cryptography, it was essentially impossible to provide key management for large-scale networks. With symmetric cryptography, as the number of users increases on a network, the number of keys required to provide secure communications among those users increases rapidly. For example, a network of 100 users would require almost 5000 keys if it used only symmetric cryptography. Doubling such a network to 200 users increases the number of keys to almost 20 000. Thus, when only using symmetric cryptography, key management quickly becomes unwieldy even for relatively small-scale networks.

The invention of public-key cryptography was of central importance to the field of cryptography and provided answers to many key management problems for large-scale networks. For all its benefits, however, public-key cryptography did not provide a comprehensive solution to the key management problem. Indeed, the possibilities brought forth by public-key cryptography heightened the need for sophisticated key management systems to answer questions such as the following:

- The encryption of a file once for a number of different people using public-key cryptography.
- The decryption of all files that were encrypted with a specific key in case the key gets lost.
- The certainty that a public key apparently originated from a specific individual is genuine and has not been forged by an imposter.
- The assurance that a public key is still trustworthy.

The next section provides an introduction to the mechanics of encryption and digital signatures.

A1.12.5.2 Encryption and digital signature explained

To better understand how cryptography is used to secure electronic communications, a good everyday analogy is the process of writing and sending a check.

- Securing the electronic equivalent of the check. The simplest electronic version of the check can be a text file, created with a word processor, asking a bank to pay someone a specific sum. However, sending this check over an electronic network poses several security problems:
- Privacy: enabling only the intended recipient to view an encrypted message. Since anyone could intercept and read the file, confidentiality is needed.
- Authentication: ensuring that entities sending the messages, receiving messages, or accessing systems are who they say they are, and have the privilege to undertake such actions. Since someone else could create a similar counterfeit file, the bank needs to authenticate that it was actually you who created the file.
- Non-repudiation: establishing the source of a message so that the sender cannot later claim that they did not send the message. Since the sender could deny creating the file, the bank needs non-repudiation.
- Content integrity: guaranteeing that messages have not been altered by another party since they were sent. Since someone could alter the file, both the sender and the bank need data integrity.
- Ease of use: ensuring that security systems can be consistently and thoroughly implemented for a wide variety of applications without unduly restricting the ability of individuals or organizations to go about their daily business.

To overcome these issues, the verification software performs a number of steps hidden behind a simple user interface. The first step is to 'sign' the check with a digital signature.

- Digital signature

    The process of digitally signing starts by taking a mathematical summary (called a *hash code*) of the check. This hash code is a uniquely-identifying digital fingerprint of the check. If even a single bit of the check changes, the hash code will dramatically change. The next step in creating a digital signature is to sign the hash code with the sender's private key. This signed hash code is then appended to the check.

    How is this a signature? Well, the recipient (in this case the bank) can verify the hash code sent to it, using the sender's public key. At the same time, a new hash code can be created from the received check and compared with the original signed hash code. If the hash codes match, then the bank has verified that the check has not been altered. The bank also knows that only the genuine originator could have sent the check because *only he has the private key that signed the original hash code.*

- Confidentiality and encryption
  Once the electronic check is digitally signed, it can be encrypted using a high-speed mathematical transformation with a key that will be used later to decrypt the document. This is often referred to as a *symmetric key* system because the same key is used at both ends of the process.
  As the check is sent over the network, it is unreadable without the key, and hence cannot be intercepted. The next challenge is to securely deliver the symmetric key to the bank.
- Public-key cryptography for delivery symmetric keys
  Public-key encryption is used to solve the problem of delivering the symmetric encryption key to the bank in a secure manner. To do so, the sender would encrypt the symmetric key using the bank's public key. Since only the bank has the corresponding private key, only the bank will be able to recover the symmetric key and decrypt the check.
  Why use this combination of public-key and symmetric cryptography? The reason is simple. Public-key cryptography is relatively slow and is only suitable for encrypting small amounts of information – such as symmetric keys. Symmetric cryptography is much faster and is suitable for encrypting large amounts of information such as files.
  Organizations must not only develop sound security measures, they must also find a way to ensure consistent compliance with them. If users find security measures cumbersome and time consuming to use, they are likely to find ways to circumvent them – thereby putting the company's intranet and extranet at risk. Organizations can ensure the consistent compliance to their security policy through:
- Systematic application. The system should automatically enforce the security policy so that security is maintained at all times.
- Ease of end-user deployment. The more transparent the system is, the easier it is for end-users to use – and the more likely they are to use it. Ideally, security policies should be built into the system, eliminating the need for users to read detailed manuals and follow elaborate procedures.
- Wide acceptance across multiple applications. The same security system should work for all applications a user is likely to employ. For example, it should be possible to use the same security system whether one wants to secure e-mail, e-commerce, server access via a browser, or remote communications over a virtual private network.

A1.12.5.3 PKI definition (public key infrastructure)

Imagine a company that wants to conduct business electronically, exchanging quotes and purchase orders with business partners over the Internet. Parties exchanging sensitive information over the Internet should always digitally sign communications so that:

- The sender can securely identify themselves – assuring business partners that the purchase order really came from the party claiming to have sent it (providing a *source authentication service*).
- An entrusted third party cannot alter the purchase orders to request hypodermic needles instead of sewing needles (*data integrity*).

If a company is concerned about keeping the nature of particulars of their business private, they may also choose to encrypt these communications (confidentiality).

The most convenient way to secure communications on the Internet is to employ public key cryptography techniques. But before doing so, the user will need to find and verify the public keys of the party with whom he or she wishes to communicate. This is where a public-key infrastructure comes in.

A1.12.5.4 PKI functions

A successful public-key infrastructure needs to perform the following:

- Certify public keys (by means of certification authorities)
- Store and distribute public keys
- Revoke public keys
- Verify public keys

Let us now look at each of these in turn.

Certification authorities

Deploying a successful public key infrastructure requires looking beyond technology. As one might imagine, when deploying a full scale PKI system, there may be dozens or hundreds of servers and routers, as well as thousands or tens of thousands of users with certificates. These certificates form the basis of trust and interoperability for the entire network. As a result, the quality, integrity, and trustworthiness of a public key infrastructure depends on the technology, infrastructure, and practices of the certificate authority who issues and manages these certificates.

Certificates authorities (CA) have several important duties. First and foremost, they must determine the policies and procedures which govern the use of certificates throughout the system.

The CA is a *'trusted third party'*, similar to a passport office, and its duties include:

- Registering and accepting applications for certificates from end-users and other entities
- Validating entities' identities and their rights to receive certificates
- Issuing certificates
- Revoking, renewing, and performing other life cycle services on certificates
- Publishing directories of valid certificates
- Publishing lists of revoked certificates
- Maintaining the strictest possible security for the CA's private key
- Ensure that the CA's own certificate is widely distributed
- Establishing trust among the members of the infrastructure
- Providing risk management

Since the quality, efficiency and integrity of any PKI depends on the CA, the trustworthiness of the CA must be beyond reproach.

On the one end of the spectrum, certain users prefer one centralized CA which controls all certificates. Whilst this would be the ideal case, the actual implementation would be a mammoth task.

At the other end of the spectrum, some parties elect not to employ a central authority for signing certificates. With no CAs, the individual parties are responsible for signing each other's certificates. If a certificate is signed by the user or by another party trusted by the user, then the certificate can be considered valid. This is sometimes called a 'web of trust' certification model. This is the model popularized by the PGP (pretty good privacy) encryption product.

Somewhere in the middle ground lies a hybrid approach that relies upon both independent CAs and peer-to-peer certification. In such an approach, businesses may act as their own CA, issuing certificates for its employees and trading partners. Alternatively, trading partners may agree to honor certificates signed by trusted third party CAs. This decentralized model most closely mimics today's typical business relationships, and it is likely the way PKI's will mature.

Building a public-key infrastructure is not an easy task. There are a lot of technical details to address – but the concept behind an effective PKI is quite simple: a PKI provides the support elements necessary to enable the use of public key cryptography. One thing is certain: the public-key infrastructure will eventually – whether directly or indirectly – reach every Internet user.

Storage and distribution of public keys

E-commerce transactions don't always involve parties who share a previously established relationship. A PKI provides a means for retrieving certificates. If provided with the identity of the person of interest, the PAKI's directory service will provide the certificate. If the validity of a certificate needs to be verified, the PAKI's certificate directory can also provide the means for obtaining the signer's certificate.

Revocation of public keys

Occasionally, certificates must be taken out of circulation, or revoked. After a period of time, a certificate will expire. In other cases, an employee may leave the company or a person may suspect that his or her private key has been compromised. In such circumstances, simply waiting for a certificate to expire is not the best option, but it is nearly impossible to physically recall all possible copies of a certificate already in circulation. To address this problem, CAs publish certificate revocation lists (CRLs) and compromised key lists KRLs).

Verification of public keys

The true value of a PAKI is that it provides all the pieces necessary to verify certificates. The certification process links public keys to individual entities, directories supply certificates as needed, and revocation mechanisms help ensure that expired or untrustworthy certificates are not used.

Certificates are verified where they are re-used, placing responsibility on all PAKI elements to keep current copies of all relevant CRLs and KRLs. In an emerging standard, on-line certificate status protocol (OCSP) servers may take on CRL/KRL tracking responsibilities and perform verification duties when asked.

VeriSign secure server IDs work in conjunction with secure sockets layer (SSL) technology, which is the standard protocol for secure, web-based communications. SSL only become functional after one installs a digital certificate, such as one's VeriSign secure server ID.

After installing one's ID, the server is able to establish SSL. SSL establishes a secure communications channel between one's server and one's customer's browser.

A1.12.5.5 Specific implementations

Cryptography is a fairly advanced subject, hence not many companies are active in this field. Four companies, however, should be noted. They are:
- VeriSign Corporation
  As mentioned earlier, a server ID also known as a digital certificate, is the electronic equivalent of a business license. Server IDs are issued by a trusted third party called a certification authority (CA). VeriSign has issued more than 1000 secure server IDs to date and their services are used by 98 of the fortune 100 companies. If the volume of sensitive transactions warrants it, a VeriSign secure server ID can be installed on a company's own intranet, enabling it to establish secure communications with any customer using a NetScape or Microsoft browser.

- Entrust Technologies
  Entrust offers services similar to those of VeriSign.
- PGP (pretty good privacy)
  PGP is a high security cryptographic application that allows people to exchange information with both privacy and authentication, and is similar in nature to the products offered by Verisign and Entrust, mentioned above, with the exception that it favors the 'web of trust' authentication modes. Another difference is that PGP is freely distributed by MIT via the WWW for non-commercial use. Unfortunately, this product can only be downloaded by American citizens or people resident in the United States.
  An interesting application of this encryption technology, is PGPfone which permits computer owners who have modems or connections to the Internet to use their computers as secure telephones by encrypting the conversation.
- Secure Computing
  Secure Computing has introduced the concept of virtual private networking (VPN). Many companies are starting to enjoy the simplicity and cost effectiveness of linking together different intranets to form a 'WAN' without conventional WAN technology. Rather, they use the Internet to interconnect the different sides, thereby creating a 'virtual private network' or VPN.
  The key to a secure 'WAN' is the fact that VPN authenticates and transparently encrypts bi-directional transmissions between intranets via the Internet, as well as to roaming users on VPN enabled laptops.

The result is a virtual wide area network, existing on the Internet infrastructure but inaccessible to any unauthorized parties.

A1.12.6 References

Internet/extranet/intranet security
General Index of Sources: http://www-ns.rutgers.edu
CERIAS (Center for Education and Research in Information Assurance and Security): http://www.cerias.com
Notes on hijack detection: http://www.netsys.com
Network monitoring (Network Flight Recorder): http://www.nfr.com
COAST (Computer Operations, Audit and Security Technology): http://www.cs-purdue.edu
ISS (Information System Support, Inc): http://www.iss-md.com
CSI (Computer Security Institute): http://www.gocsi.com
Network Security Policies: http://www.baselinesoft.com
CERT Coordination Center (Carnegie Mellon Software Engineering Institute): http://www.cert.org
Internet Security Magazine: http://www.securecomputing.com
An example of specific product updates e.g. MicroSoft Office: http://officeupdate.microsoft.*com*

Encryption
Secure Computing: http://www.sctc.com
VeriSign: http://www.verisign.com
PGP: http://pgp5.mit.edu
Entrust: http://www.entrust.com

Firewalls, proxy servers etc
CISCO Systems: http://www.cisco.com
SECURE Computing: http://securecomputing.com
Security Operating Systems

Virtual private networking
Firewall Report Overview: http://www.outlink.com
Secure Zone: http://www.sctc.com
WinGate : http://www.wingate.com

Index

Advanced Research Projects Agency Network (ARPANET), 87–90, 168
ALSTOM:
  products, 156–63
    bay units, 157–8
    MiCOM S10, 158–9
    MiCOM S100, 160–1
    overview, 156
    protection, 156–7
    PSCN 3020, 161–2
    SPACE 2000, 162–3
American Department of Defense (DoD), 87
Application processes (APs), 120
Application protocol control information (APCI), 98
Application protocol data unit (APDU), 98
Architectures:
  types, 128–35
    Type 1 systems, 128–30
    Type 2 systems, 130–1
    Type 3 systems, 132–3
    Type 4 systems, 133–5
Asea Brown Boveri (ABB):
  products overview, 141
  Type 1 products, 141
    communication, 142–3
    control function, 142
    measurement, 142
    monitoring, 142
    protection, 141–2
    SCADA, 143
Automation industry development, 10–12

Battery-tripping unit (BTU), 22–3
  full back-up supply, 23
  incorporation of supervision relay, 23
Bay controllers, 47
Berners-Lee, Tim, 168
Bolt, Beranek and Newman Inc (BBN), 87, 168
Busbar protection, 36, 41–2
  arc protection, 42
  blocking, 43–4
  zone protection, 42

Carrier sense multiple access with collision detection (CSMA/CD), 59–60
CGI *see* Common gateway interface (CGI)
Common gateway interface (CGI), 170
Communication, 4–5, 13
  ABB products, 142–3
  basic requirements:
    agreed medium, 51
    common context, 51
    physical link, 51
    receiver identified, 51
    same language, 51
  digital data, 51–66
  digital transmission, 51
  GE products, 140
  power system automation requirements, 119–27
  protocols, 67
    DNP V3.0, 67–78
    Ethernet, 84–6
    IEC 60870-5-101, 98–102
    IEC 60870-5-103, 102–106
    LonTalk, 82–3

modbus, 78–9
modbus plus, 79–82
PROFIBUS, 93–8
standardization, 106–107
TCP/IP, 87–93
UCA2.0, 106
requirements, 119, 121–7
SCADA, 50, 120–1
SEL products, 145–6
Computer Emergency Response Team (CERT), 181
Computer industry, 12–13
Cryptography *see* Public key infrastructure (PKI)
CSMA/CD *see* Carrier sense multiple access with collision detection (CSMA/CD)
Current transformers (CTs), 4
metering, 24–5
protection, 25–6

Data communication *see* Digital data communication
De-militarized zones (DMZs), 181
Digital data communication, 51–2
media access control principles, 55
conventional polling, 56–7
CSMA/CD, 59–60
polling by exception, 57
time-division-multiplex, 57–8
token passing, 58–9
OSI model:
basic principles of transmitting data, 60–1
seven-layer stack, 61–2
using the stack, 62–4
performance criteria:
bandwidth, 65
data throughput, 65
error rate, 65
response time, 65–6
signal-to-noise ratio, 65
transmission speed, 64–5
techniques:
master-slave, 55
peer-to-peer, 55
topologies:
bus, 54–5
defined, 52
ring, 53–4
star, 52–3
Digital technology:
automation industry, 10–12
computer industry, 12–13
electrical protection, 8–9
Distributed control systems (DCS), 10, 18
Distributed network protocol (DNP V3.0), 67–8
data link layer, 72
frame outline, 72–3
function codes, 73

request-respond, 76–7
reset of secondary link, 73–4
reset of user process, 74
send-confirm user data, 74
send-NACK, 76
send-no reply expected, 75
transmission procedures, 73–7
development, 69
IEC/IEEE, 68
interoperability, 68
open standard, 68
physical layer, 69
dial-up modem, 72
four-wire multi-point, 71–2
four-wire point-to-point, 71
modes, 70–2
topologies, 69
two-wire multidrop, 70
two-wire point-to-point, 70
SCADA, 68
transport layer (pseudo-transport), 77
application layer, 78
conclusion, 78
Domain name service (DNS), 180

Earth fault:
phase-to-earth, 34–5
restricted, 35
sensitive/sustained, 35
settings, 35
Electric Power Research Institute (EPRI), 106
Electrical protection:
arc resistance, 21
communication network, 121–2
higher level, 123–4
lower level, 122–3
comparisons of electromechanical/digital relays, 44–5
components:
fuses, 21–2
instrument transformers, 23–6
relay/circuit-breaker combination, 22–3
described, 20
development, 7
digital relays, 8–9
electromechanical relays, 7–8
microprocessor relays, 9
multi-functional, 9
static relays, 8
main functions of relays:
differential protection, 35–7
distance protection, 38
earth fault, 34–5
fault current, 33–4
frequency regulation, 38

Electrical protection (*Continued*)
    negative sequence protection, 38
    overcurrent, 32–4
    overload, 33
    voltage regulation, 38
  qualities, 26
    current differential, 28
    discrimination/selectivity, 28–32
    non-unit, 29
    overlapping zones, 26–7
    reliability, 32
    sensitivity, 32
    speed of operation, 32
    stability, 32
    unit, 28–9
    zone, 28
  specific applications:
    busbar protection, 41–4
    feeder protection, 40–1
    general protection, 39–40
    overview, 38–9
    transformer protection, 40
  symmetrical faults, 20–1
  unbalanced/asymmetrical faults, 21
Electromechanical relays:
  compared with digital relays, 44–5
  current differential relays, 7
  inverse definite minimum time (IDMT), 7
  maintenance/reliability, 8
  thermal overcurrent relays, 7
Electronic industry development:
  digital technology, 9–10
  transistor, 9
Encryption *see* Public key infrastructure (PKI)
Enhanced performance architecture (EPA), 64, 98
Enhanced RTU (ERTU), 47
Equipment installation:
  competence management, 165
  constraints, 165
  electrical protection, 165–6
  guidelines, 167
  justification, 164–5
  power system automation and Internet, 167
  suppliers, 167
Ethernet, 106, 149
  CSMA/CD access method, 84
  development, 84
  medium access control, 85–6
  signaling method, 84–5
  standards, 84
European Organization for Nuclear Research (CERN), 168
eXtensible Markup Language (XML), 171

Fieldbus message specification (FMS), 94
Firewalls, 180
  application layer (proxy servers), 182
  DMZs *see* De-militarized zones (DMZs)
  dynamic IP address allocation, 180–1
  network layer, 180
  strike back intruder response, 181
  TCP sequence number randomization, 181

Gas insulated switchgear (GIS), 10, 23
General Electric (GE):
  products, 137–41
    communications, 140
    control facilities, 139–40
    metering, 140
    monitoring, 140
    overview, 136–7
    protection features, 137–9
    SCADA, 141
Generator protection, 39–40
  electrical faults, 40
  loss of excitation function, 40
  maintain magnetic field in stator, 40
  reverse power, 40

HTML *see* Hypertext Markup Language (HTML)
HTTP *see* Hypertext Transfer Protocol (HTTP)
Human–machine interface (HMI), 112, 113–15
Hypertext Markup Language (HTML), 169
Hypertext Transfer Protocol (HTTP), 169–70

IEC 60870-5-101 protocol:
  application functions, 100–101
    acquisition of events, 100
    background scan, 100
    command transmission, 101
    cycle transmission, 100
    file transfer, 101
    station initialization/interrogation, 100
    transmission of integrated totals, 101
  application layer, 99
  application service data units, clock synchronization, 101–102
  frame integrity, 9
  overview, 98–9
  physical configuration, 9
IEC 60870-5-103 protocol:
  overview, 102
  standard information numbers, 103
    auto-reclosure indications in monitor direction, 104
    basic application functions, 106
    earth fault in monitor direction, 104
    fault in monitor direction, 104
    general commands in control direction, 105

generic functions in control
    direction, 105–106
generic functions in monitor
    direction, 105
measurands in monitor direction, 105
status indications in monitor direction, 103
supervision indications in monitor
    direction, 103
system functions in control direction, 105
system functions in monitor direction, 103
Institute of Electrical and Electronic Engineers
    (IEEE), 106
Intelligent electronic device (IED), 4, 9
    communications, 50
    control, 49
    definition, 48
    functions, 48
    metering, 50
    monitoring, 49–50
    protection, 48–9
International Standards Organization (ISO), 91–2
Internet, 167
    communications, 175
        hardware/software, 175
        speed/bandwidth, 175–6
    connecting multiple hosts:
        via dedicated router, 175
        via IP sharer, 174
        via NAT server (IP masquerading), 174
        via proxy server, 174
        via UNIX/NT gateway, 174–5
    connecting remote hosts to corporate
        LAN, 173
    connecting single host, 172–3
    history/background, 168–71
    layers, 92
        host-to-host, 93
        Internet, 92
        network interface, 92
        network management, 93
        process/application, 93
    security:
        firewalls, 180–2
        IDS, 182–7
        overview, 176–8
        passwords, 178
        PKI, 187–92
        routers, 178–80
Internet protocol (IP) see TCP/IP
Internet service providers (ISPs):
    access methods, 172–3
        cellular (mobile) phone, 173
        dial-up modem, 172–3
        ISDN connection, 173
        leased lines, 173
        primary, 172
        secondary, 172
Intrusion detection systems (IDS), 182
    host-based, 182–3
    network flight recorder, 183
    network-based, 183

Java, 170
JavaScript, 170–1

Link protocol control information
    (LPCI), 98
Link protocol data unit (LPDU), 98
Local area networks (LANs), 12,
    109, 173
Local intelligence see Intelligent electronic
    device (IED)
LonTalk protocol, 82
    features, 82
        addressing, 83
        communication rates, 83
        multiple communication channels, 83
        multiple media support, 83

Measurement, 3, 13
    ABB products, 142
    higher level, 126
    lower level, 125–6
Modbus plus protocol, 79
    characteristics, 80
    error checking/recovery, 81
    how network works, 80
    how nodes access network, 81
    network terminology, 80
    peer cop transactions, 81–2
    physical network, 80
Modbus protocol:
    functions, 79
    overview, 78–9
Monitoring, 126
    ABB products, 142
    example, 127
    GE products, 140
    higher level, 127
    lower level, 127
    SEL products, 145
Motor control center (MCC), 15, 18

Natural address translation (NAT), 180

Open system interconnection (OSI)
    model, 60–1, 87–90
    construction, 61–2
        application layer, 62
        data link layer, 62

Open system interconnection (OSI) (*Continued*)
    network layer, 62
    physical layer, 62
    presentation layer, 62
    session layer, 62
    transport layer, 62
  enhanced protocol architecture, 64
  point-to-point connections, 64
  standards, 91–2
  using the stack, 62–4
Open system interconnection – reference model (OSI-RM), 87, 89–90
Overcurrents:
  extremely inverse (EI), 33
  fault current, 33–4
  inverse definite minimum time (IDMT), 32
  normally inverse (NI), 32–3
  overload, 33
  very inverse (VI), 32–3

Perl *see* Practical Extraction and Report Language (Perl)
Personal computers (PCs), 108, 110–11
Power networks:
  described, 15–16
  distribution, 15, 18
  generation, 15, 16
  low voltage applications, 18
  network studies, 18–19
  position of substation, 16
  transmission, 15, 16–18
  typical example, 14
Power system automation:
  architecture, 4
    communications network, 4–5
    object division, 4
    SCADA master, 5
  communication requirements, 119
    control, 124–5
    example, 127
    measurements, 125–6
    monitoring information, 126–7
    protection, 121–4
  configuration, 120
  definition, 1
  development, 7
    automation industry, 10–12
    communications industry, 13
    computer industry, 12–13
    electrical industry, 7–9
    electronic industry, 9–10
    measurement industry, 13
    switchgear industry, 10

  main components, 1
    control, 2–3
    data communication, 3
    electrical protection, 2
    measurement, 3
    monitoring, 3
  systems on market *see* ALSTOM; Asea Brown Boveri; General Electric; SEL; Siemens
Practical Extraction and Report Language (Perl), 171
PROFIBUS:
  application profiles, 95
  broadcast communication, 98
  communication profiles, 94
    DP, 94
    FMS, 94
  multicast communication, 98
  overview, 93
  physical profiles, 94
    IEC 1158-2, 95
    RS-485, 94–5
  protocol architecture, 95–8
Programmable logic controllers (PLCs), 4, 10–12, 18, 47, 108
Protection relays, 47–8
Protocols *see* Communication protocols
Public key infrastructure (PKI):
  cryptography, 187–8
  definition, 189–90
  encryption/digital signature, 188–9
  functions, 190
    certification authorities, 190–1
    revocation, 191
    storage/distribution, 191
    verification, 191
  specific implementations, 191–2

Remote substation access, 46–50
Remote terminal units (RTUs), 4, 10–12, 46–7, 108
Routers, 178–9
  protecting, 179
    allow Telnet access, 179
    authenticate protocol, 179
    current software, 179
    release notes/vendor bulletins, 179
    SNMP, 179
    verify manager password, 179
  protecting network, 179–80
    access control lists (ACLs), 179–80
    logging, 179

SCADA *see* Supervisory control and data acquisition (SCADA)
  ABB products, 143
  communications, 50, 120–1

definition/background, 108–109
expandability of system, 118
GE products, 141
general features:
  access to data, 117
  alarms, 115–16
  client/server distributed
    processing, 117
  IED interface, 116
  networking, 10–11, 117
  trends, 116
hardware, 109–11
IEDs, 48
requirements, 109–15
RTUs, 46–7
SEL products, 146
Siemens product, 152–3
software, 111
  archiving/database storage, 112
  control, 112
  data acquisition, 112
  human–machine interface, 113–14
  operating system, 111
  packages, 12–13
  SCADA application software, 115
  SCADA system software, 111–12
system response times, 117–18
Scripting languages, 170–1
Security management:
  advisory services, 186–7
  certification, 183–4
    access to web, 184
    basic knowledge of TCP/IP, 184
    management directive, 184
    security policy, 184
    time, 184–5
  information policies, 185–6
SEL:
  future developments, 148–9
    Ethernet, 149
    SEL-2701/utility communications
      architecture, 149
    serial, 148
  products:
    advantages, 146–8
    control, 145
    data communications, 145–6
    disadvantages, 148
    metering, 145
    monitoring, 145
    overview, 143–4
    protection, 144–5
    SCADA, 146
  suppliers information, 146–9
Server, 171
Siemens:
  products, 150–6
    protection, 150
  SICAM PCC, 153
    advantages, 155
  SICAM RTU, 154–5
  SICAM SAS, 154
    disadvantages, 156
  SINAUT LSA, 150–1
    compact bay control unit 6MB524, 152
    compact remote terminal unit 6MB552, 152
    input/output unit 6MB520, 152
    minicompact remote terminal
      unit 6MB5530-0, 152
    SCADA, 152–3
    station control unit 6MB5510/5515,
      151–2
Smart RTU, 47
Stanford Research Institute, 87
Substation control:
  communication network, 124
  higher level, 124–5
  lower level, 124
Supervisory control and data acquisition (SCADA), 2, 3, 4, 5, 22
Switchgear, 10
  fault protection, 21

TCP/IP, 168, 181
  Internet layers, 92–3
  model, 87–90
  origin/background, 87
  standards, 91–2
Time-division-multiplex media access
  (TDM), 57–8
Transport control protocol (TCP) *see* TCP/IP

Utilities Communication Architecture, Version 2
  (UCA 2.0), 106

Voltage transformers (VTs), 4
  electromagnetic and capacitive, 24

Wide area networks (WANs), 12, 109

Xerox Networking System (XNS), 87
XML *see* eXtensible Markup Language (XML)

THIS BOOK WAS DEVELOPED BY IDC TECHNOLOGIES

WHO ARE WE?

IDC Technologies is internationally acknowledged as the premier provider of practical, technical training for engineers and technicians.

We specialise in the fields of electrical systems, industrial data communications, telecommunications, automation & control, mechanical engineering, chemical and civil engineering, and are continually adding to our portfolio of over 60 different workshops. Our instructors are highly respected in their fields of expertise and in the last ten years have trained over 50,000 engineers, scientists and technicians.

With offices conveniently located worldwide, IDC Technologies has an enthusiastic team of professional engineers, technicians and support staff who are committed to providing the highest quality of training and consultancy.

TECHNICAL WORKSHOPS

TRAINING THAT WORKS

We deliver engineering and technology training that will maximise your business goals. In today's competitive environment, you require training that will help you and your organisation to achieve its goals and produce a large return on investment. With our "Training that Works" objective you and your organisation will:

- Get job-related skills that you need to achieve your business goals
- Improve the operation and design of your equipment and plant
- Improve your troubleshooting abilities
- Sharpen your competitive edge
- Boost morale and retain valuable staff
- Save time and money

EXPERT INSTRUCTORS

We search the world for good quality instructors who have three outstanding attributes:

1. Expert knowledge and experience – of the course topic
2. Superb training abilities – to ensure the know-how is transferred effectively and quickly to you in a practical hands-on way
3. Listening skills – they listen carefully to the needs of the participants and want to ensure that you benefit from the experience

Each and every instructor is evaluated by the delegates and we assess the presentation after each class to ensure that the instructor stays on track in presenting outstanding courses.

HANDS-ON APPROACH TO TRAINING

All IDC Technologies workshops include practical, hands-on sessions where the delegates are given the opportunity to apply in practice the theory they have learnt.

REFERENCE MATERIALS

A fully illustrated workshop book with hundreds of pages of tables, charts, figures and handy hints, plus considerable reference material is provided FREE of charge to each delegate.

ACCREDITATION AND CONTINUING EDUCATION

Satisfactory completion of all IDC workshops satisfies the requirements of the International Association for Continuing Education and Training for the award of 1.4 Continuing Education Units.

IDC workshops also satisfy criteria for Continuing Professional Development according to the requirements of the Institution of Electrical Engineers and Institution of Measurement and Control in the UK, Institution of Engineers in Australia, Institution of Engineers New Zealand, and others.

CERTIFICATE OF ATTENDANCE

Each delegate receives a Certificate of Attendance documenting their experience.

100% MONEY BACK GUARANTEE

IDC Technologies' engineers have put considerable time and experience into ensuring that you gain maximum value from each workshop. If by lunch time of the first day you decide that the workshop is not appropriate for your requirements, please let us know so that we can arrange a 100% refund of your fee.

ONSITE WORKSHOPS

All IDC Technologies Training Workshops are available on an on-site basis, presented at the venue of your choice, saving delegates travel time and expenses, thus providing your company with even greater savings.

OFFICE LOCATIONS

AUSTRALIA • CANADA • IRELAND • NEW ZEALAND • SINGAPORE • SOUTH AFRICA • UNITED KINGDOM • UNITED STATES

idc@idc-online.com • www.idc-online.com

Visit our Website for FREE Pocket Guides

IDC Technologies produce a set of 4 Pocket Guides used by thousands of engineers and technicians worldwide.

- Vol. 1 - ELECTRONICS
- Vol. 2 - ELECTRICAL
- Vol. 3 - COMMUNICATIONS
- Vol. 4 - INSTRUMENTATION

To download a **FREE copy** of these internationally best selling pocket guides go to:
www.idc-online.com/freedownload/

8450046